Making Things Stick

Making Things Stick

Surveillance Technologies and Mexico's War on Crime

Keith Guzik

UNIVERSITY OF CALIFORNIA PRESS

University of California Press, one of the most
distinguished university presses in the United States,
enriches lives around the world by advancing scholarship
in the humanities, social sciences, and natural sciences.
Its activities are supported by the UC Press Foundation
and by philanthropic contributions from individuals and
institutions. For more information, visit www.ucpress.edu.

University of California Press
Oakland, California

Suggested citation: Guzik, Keith. *Making Things Stick:
Surveillance Technologies and Mexico's War on Crime.*
Oakland: University of California Press, 2016.
doi: http://dx.doi.org/10.1525/luminos.12

Library of Congress Cataloging-in-Publication Data

Names: Guzik, Keith, author.
Title: Making things stick : surveillance technologies and Mexico's war on
   crime / Keith Guzik.
Description: Oakland, California : University of California Press, [2016] |
   "2016 | Includes bibliographical references and index.
Identifiers: LCCN 2015040252| ISBN 9780520284043
(pbk. : alk. paper) | ISBN 9780520959705 (electronic)
Subjects: LCSH: Crime prevention—Mexico. | Social control—Government
   policy—Mexico. | Electronic surveillance—Mexico. | Security
   systems—Mexico.
Classification: LCC HV7434.M6 G89 2016 | DDC 363.2/32—dc23
LC record available at http://lccn.loc.gov/2015040252

Анечке

Contents

Illustrations

FIGURES

Acknowledgments

This book is made possible only through the generous support of numerous organizations and individuals. First among these is the National Science Foundation, which funded my research with a grant (no. 1024469) co-awarded through its Science, Technology, and Society and Law and Social Science programs. No less central were state institutions and employees in Mexico, including administrators and frontline workers with the Public Registry of Vehicles in Zacatecas and Sonora and their federal counterparts in the Mexico City offices of the Executive Secretary of the National System for Public Security. Their cooperation was remarkable, and this work would not have been completed without it. I owe my thanks as well to the car companies in Mexico that made their representatives available for interviews.

The institutions of higher education where I have worked over the past decade were essential to this work too. The administration and faculty at Bloomfield College supported my research through a study leave in spring 2011 that enabled me to complete the majority of my fieldwork, and they also invited me to present preliminary findings at faculty forums. Beyond being supportive friends, the faculty at the college were an enduring example of striking the proper balance between research, teaching, and service in the academy. The University of Colorado, Denver, meanwhile, provided both a nurturing work environment that facilitated writing this book and funding to help publish it as an open access book. The CU Denver Department of Sociology gave generous funding as well for a research assistant, who was critical in helping me clear the final hurdles of the research process. And my new colleagues in the department and the College of Liberal Arts and Sciences were wonderfully supportive in welcoming me to my new home in the Rocky Mountains.

I also want to thank the University of California Press, and Maura Roessner and Jack Young in particular. Maura is an excellent editor to work with. And she made navigating the uncharted terrain of open access publishing an exciting and enjoyable experience. I am indebted as well to Julie Van Pelt, my copyeditor, who did wonderful work in improving the book's readability.

Outside the organizations that supported this work, a number of people deserve acknowledgment. In Mexico, Daiset Ruiz-Sarquis, Armando López Muñoz, and Carmen Cebreros Urzaiz provided critical insight into the endlessly rich history and culture of the country that would have been impossible for me to uncover on my own. But more than this, they offered immeasurable warmth and a good dose of madness that helped carry me through the drudgiest days of field research. I owe thanks as well to Damián Ortega, Betsabeé Romero, Margarita Cabrera, Kurimanzutto, and the Asociación de Amigos del Museo del Arte Popular for allowing me to use images of their artwork in this book.

Two research assistants contributed immensely to this project. At Bloomfield College, I was lucky to cross paths with Nora López Matta, who skillfully transcribed interviews and coded survey data while balancing the financial and familial demands of pursuing her American dream. At the University of Colorado, Denver, Heather Worrell generously offered the help of her Spread the Word language services company to help edit transcriptions and create a nimble digital bibliography from the messy mass of sources that I used to put this book together.

The ideas that eventually resulted from this research were improved through the kind, critical feedback of colleagues. Special recognition is owed to those who were able to trudge through earlier versions of the manuscript at the request of UC Press—Diane Davis, Katja Franko Aas, and other anonymous reviewers. I am indebted as well to Gary T. Marx and Robert Buffington, who selflessly offered insightful comments on the whole manuscript. I also owe thanks to those who provided me venues for presenting preliminary findings at professional conferences, including Nicholas Rowland and Jan-Hendrick Passoth and their work group on science and technology studies (STS) and the state; and Karen

Levy and Aaron Smyth and their research network on the intersections of STS and sociolegal studies. A number of others—Jon Gilliom, Mary Mitchell, William Rose, Margaret Hu, Bryce Newell, Diana Mincyte, and Andrzej Nowak—provided helpful comments on conference papers and presentations that form the basis of this book's chapters. Finally, I want to especially thank Anna Maria Marshall and Evan Stark. Their areas of specialization may lie outside this work, but their continued mentorship well past the time when one should require it is a gift I will always appreciate.

CHAPTER 1

Surveillance Technologies and States of Security

1.1 BUNKER MENTALITY

Mexico's Federal Police Intelligence Center (CIPF) was inaugurated on November 24, 2009, in a ceremony attended by President Felipe Calderón and Secretary of Public Security Genaro García Luna. The CIPF, a subterranean structure colloquially known as El Bunker, serves as the command center for the federal government's War on Crime.[a] It houses Plataforma México, a network of advanced telecommunication and information technologies receiving data from over six hundred state and municipal offices; 169 federal police stations; the national registries of people, vehicles, criminal records, fingerprints, and ballistics; and video cameras located throughout the country, including those at the Basilica of Our Lady of Guadalupe, dedicated to the Virgin of Guadalupe, the patron saint of Mexico.[1] To visualize the data, El Bunker features four video walls, each measuring 65 by 10 feet, displaying eighty rear-screen projectors arranged in four 2 by 10 configurations.[2] In his remarks, President Calderón claimed that the center would serve as a

a. The federal government's security campaign has carried various monikers—la Guerra contra el Narco (the War against the Narcos), la Lucha contra la Inseguridad (the Fight against Insecurity), la Guerra contra el Crimen (the War against Crime or the War on Crime), among others. Of these, the War against the Narcos and the War on Crime were the most common during the Calderón administration. I use the term "War on Crime" because it captures the fact that the government is targeting forms of illicit behavior beyond drug trafficking.

"computer brain" to keep the federal police "a step ahead of crime." It would allow Mexico to win its War on Crime, he explained, since "wars are won with this, with technology, information, intelligence, planning, and force."[3] With these words, the Mexican leader gave voice to his administration's faith in the power of technology to defeat crime.

Situated off the southwest corner of Chapultepec Park, Mexico City's verdant oasis, El Bunker's proximity to the park's other iconic buildings—Chapultepec Castle and Los Pinos—provides a commentary on the evolving relationship between governors and the governed in Mexico. Chapultepec Castle, located at the highest point of the park, is a regal structure that was commissioned by Viceroy Bernardo de Gálvez in 1775 and given its current appearance by the Austrian-born Emperor Maximilian I of Mexico in 1864. Behind the castle in the south-central section of the park, the more austere Los Pinos was constructed on the order of President Venustiano Carranza in 1917 and has been the primary home for Mexico's heads of state since 1934, when President Lázaro Cárdenas moved the presidential residence out of Chapultepec Castle. El Bunker, meanwhile, with its central conference room that seats the president and his security cabinet in the event of a national emergency, is a two-story underground structure powered by an independent energy source. If Chapultepec Castle pronounced the presence of a royal authority through its privileged position above Mexico City, and if Los Pinos symbolized the progressive ideals of Mexico's postrevolutionary government to level the distance between the country's most and least powerful sectors, El Bunker embodies the security anxieties of the contemporary government, which would secure society by placing its administrative center outside the grasp of the general population while maintaining oversight through advanced surveillance technologies.

A solid five years into the intelligence center's existence, its value remains in doubt. Although homicides, robberies, and extortions are down in recent years, violent crime remains high throughout the country.[4] And the government's limited capacity to combat criminal wrongdoing has been underscored by dramatic events such as the massacre of forty-three students of the Raúl Isidro Burgos Rural Teachers' College of Ayotzinapa in September 2014 and the escape of Joaquín "El Chapo" (Shorty) Guzmán from a maximum security prison in July 2015. These high-profile crimes, like most delinquency in Mexico, remain unsolved. As regards the center itself, it has been dogged by numerous problems, including unmanageable historical data, unreliable interagency communications, the reluctance of state agencies to share data, and manual processes of information keeping at the local level that slow data processing and accuracy.[5] These challenges have not diminished the government's faith in technology. "It is a matter of time," officials assert, when asked about the center's impact on crime.[6] And additional police bunkers have been constructed in Mexico since.[7] But in Mexico's War on Crime, one wonders whether time and technology will be enough.

1.2 LIVING IN THE SURVEILLANCE SOCIETY

This is a book about surveillance technologies and their impact on the relationship between authorities and those they govern. Surveillance, defined as "any collection or processing of personal data, whether identifiable or not, for the purposes of influencing or managing those whose data have been garnered,"[8] has become a topic of growing popular and scholarly interest the last twenty years. This is reflected in the attention paid to it by Hollywood (in films such as *The Truman Show, Gattaca, Lost Highway, Minority Report, Panic Room*) and academia (in the journal *Surveillance and Society,* as well as a number of recent books). This growing popularity does not owe to the novelty of surveillance, since surveillance is not new. Political authorities have always kept track of people, just as parents have always looked after their children, teachers tracked their students, doctors monitored their patients, and bosses watched over their workers.

But how we watch has changed, thanks to the proliferation of computers, mobile devices, CCTV cameras, RFID chips, and other gadgets in society today. "Traditional surveillance," the seminal surveillance scholar Gary T. Marx notes, was characterized by "close observation, especially of a suspected person." "New surveillance," however, is performed "through the use of technical means to extract or create personal or group data, whether from individuals or contexts."[9] And on a daily basis, we come into contact with a host of technologies whose surveillant capacities are transforming the contours of social life. "Helicopter parents" wield "electronic leashes" to remain ever present in the lives of their children, classrooms are turned "inside out" or made into "MOOCs" to accommodate greater numbers of students, doctors connect to patients in "eICUs," and "job spill" and "workweek creep" befall greater numbers of workers.

At the level of national security, networks of computers armed with powerful processors and sophisticated software scoop up data transmitted across the Internet and allow governments to track and store the content of people's digital communications. Backscatter X-ray imaging devices enable security officers at airports to effortlessly photograph travelers through their clothes. The advent of unmanned aerial vehicles (UAV) armed with missiles that can be guided by video operators has simplified the assassination of suspected terrorists overseas. Closed-circuit television (CCTV) and Internet-protocol television cameras allow dummy police officers to monitor public spaces. Biometric technologies such as fingerprinting and iris scans make possible the identification of billions of people across the globe. And tracking cars and people would be many times more difficult without the radio-frequency identification (RFID) chips or global positioning system (GPS) devices that can attach to them.

The ubiquity of surveillance technologies might make us yearn for earlier times, when our lives were not shackled by such objects. But we should resist such knee-jerk reactions. The sense that privacy is under assault today is not imagined. The unsolicited email or phone call speaks to the capacity of information systems to link diverse databases.[10] The ability of online vendors such as Amazon or Netflix to predict our reading, listening, or viewing preferences evidences how those watching know more about us than we do ourselves.[11] And the requirement that we possess a driver's license in order to board an airplane speaks to the propensity of technologies to not only "creep"[12] but "surge"[13] into applications beyond their original design. But we must weigh these concerns against considerations of how privacy and personal data are not so much invaded by surveillance technologies as willingly offered up by people as "tokens of trust"[14] necessary for social exchanges in today's world. We share the details of our lives on Facebook or with online vendors so that we can move more freely and buy more cheaply than if constrained to our immediate community. Loss of privacy thus equates with greater freedom of movement and other privileges. In addition, technologies such as cell phones and encryption programs enhance privacy by providing greater anonymity than in the past.[15] Online purchasing is more or less secure. And encryption and anonymization programs such as GnuPG and Tor have proven to be effective enough that the US government purposefully retains encrypted or anonymous data in the hopes of cracking their codes.[16]

Privacy concerns connect with identity as well. The digital identities that people construct on Facebook (or have constructed for them based on the digital trails they leave online) serve as "data doubles,"[17] "dividuals,"[18] or "electronic doppelgängers."[19] At the same time, biometric technologies such as iris scans and genetic fingerprinting assign identity by anchoring it to one's body.[20] In both instances, technologies threaten the self,[21] as people find their ability to define who they are reduced and the line between the self and other fades.[22] The ubiquity of surveillance technologies affects group identity too. CCTV cameras installed in housing developments disrupt forms of community as people cut the amount of time spent in open areas where community engagement traditionally occurs.[23] That said, biometric technologies such as those utilized in India's population registry hold the possibility of guaranteeing personal identity and the civic rights and privileges that citizenship entails.[24] Likewise, DNA testing has revolutionized the criminal justice system in the United States,[25] helping identify scores of individuals wrongfully incarcerated for crimes they did not commit and thus bringing some measure of justice to individuals and communities wrongfully targeted by law enforcement.

DNA tests bring to mind the reliability of technical methods of seeing and their potential to predict crime[26] and realize "front end control."[27] There is evidence that the innovative use of crime statistics in programs such as COMPSTAT is effective in reducing property crime.[28] And government officials cite the efficacy of surveillance in stopping terrorist attacks.[29] But it is of course impossible to know that such malicious plans would not have been thwarted by more conventional law-enforcement tactics. What is more, these technologies were unable to preempt the Boston Marathon bombers, Dzhokhar and Tamerlan Tsarnaev, even though surveillance programs "discriminate by design" against foreign nationals.[30] Also, with relation to ordinary street crime, organizational variables impinge on the performance of COMPSTAT[31]—police departments continue to define policing as patrol work and do not undertake the institutional reorganization called for under the program. Surveillance technologies are additionally challenged by the uncertain nature of the phenomena they look to control. The spelling of names[32] and the features of faces[33] change over time, and data can be coded in incompatible formats,[34] duplicate places,[35] or stripped of meaning in the coding process.[36]

The uneven performance record of surveillance technologies in providing security might give public officials pause in adopting them. But this has clearly not been the case.[37] At the time of writing, local, state, and federal law-enforcement agencies and local legislatures in the United States are exploring the adoption of UAVs or drones to fight crime[38] and combat illegal immigration.[39] This disconnect between actions and outcomes speaks to, on the one hand, the influence of "security cultures," or "prevailing understandings of threats and appropriate responses to them," on public policy in the United States.[40] TV shows, movies, and news reports disproportionately focus on terrorism and glamorize the ability of sophisticated technologies and muscular men using extralegal means (James Bond, Ethan Hunt, Jack Bauer) to stop terrorism.[41] The September 11, 2001, terrorist attacks in the United States also intensified existing trends of "governing through crime,"[42] where authorities use crime prevention as a rationale for expanding techniques and technologies of social control. The disconnect between the government's increasing adoption of technologies and the lack of evidence for their efficacy also speaks to, on the other hand, the influence of the private sector over public life. Private companies produce and operate many of these technologies.[43] Companies lobby governments to adopt their wares[44] and advertise so that individual citizens adopt them to make themselves safe. In the United Kingdom, private CCTV operators gradually assume discretionary power from police officers in deciding whom to place under watch.[45] In this sense, instead of Big Brother, it makes more sense today to speak of the increasing number of Little Brothers who exercise authority over ordinary people.[46]

The growing influence exercised by private actors gives the impression that fundamental processes affecting the nature of political life in our democratic societies are increasingly out of our hands. Programs of surveillance and secret prisons for terrorist suspects operate outside the US court system and beyond public oversight.[47] At the theoretical level, many scholars have observed that surveillance technologies are part of an alteration in the nature of power in society underscored by "social sorting."[48] At national borders, programs like US VISIT sort between safe/legitimate and dangerous/illegitimate travelers.[49] On city streets, CCTV operators differentiate between desirables and undesirables, and intelligent transportation systems (ITS) implement "throughput rationality" to prioritize certain mobilities (motor vehicles) over others (pedestrians).[50] In correctional facilities, good risks for rehabilitation are separated from bad risks.[51] Slowly, political philosophies enshrining universal rights are yielding to utilitarian technocratic practices dividing us between good and bad risks.

For many, this ongoing advance of surveillance technologies engenders a breakdown in "institutional trust" and invites resistance.[52] Resistance can run from the momentous to the mundane. Concerned individuals, such as Chelsea Manning and Edward Snowden, with knowledge of or access to secret government communications and programs have leaked information to the public. Concerned organizations, such as the Electronic Frontier Foundation (EFF) or Electronic Privacy Information Center (EPIC), have brought lawsuits against the US government and communication companies for violating privacy laws and civil liberties.[53] Meanwhile, individuals confronting drug testing at work might either directly refuse to participate in testing, avoid it by hiding or not attending work, switch clean for tainted samples, distort the test by consuming substances to neutralize the evaluation, and so on.[54] Those collecting welfare benefits will engage in paid labor in violation of state regulations.[55] And people under the watch of CCTV monitoring take to playing for the camera[56] or defy the authority of watchers by displaying a middle finger.[57]

1.3 SURVEILLANCE TECHNOLOGIES IN AN INSECURE WORLD

But how would we react to surveillance if we found ourselves elsewhere? Globally, authorities are exercising novel means for monitoring people in the name of security. France operates a massive secret telecommunications surveillance program designed to identify security threats.[58] Before the United States and France, the government of Nigeria was suspected of operating a program to monitor Internet communications.[59] India is creating a national registry of people based on the fingerprints and iris scans of its over one billion citizens.[60] Thailand has begun installing video cameras into the life-sized, fiberglass decoy police officers that dot the country in order to combat street crime.[61] Brazil has created an electronic vehicle registration program that will be able to track motor vehicles by attaching RFID chips to them.[62] And the list goes on.

If there are now "eyes everywhere"[63] and surveillance technologies are an essential aspect of globalization, this does not mean that they always see in the same way. Local, national, and regional contexts shape their distribution and intensity.[64] In Rio de Janeiro, video cameras are employed by competing police forces from different levels of government and by private citizens in pursuit of security against organized crime, while in Tokyo they are embraced by civic community groups in a highly bureaucratized fashion that is indicative of a "surveillance society."[65] Within Latin America, countries such as Brazil, Mexico, and

Venezuela are outpacing others in their turn to surveillance technologies to ensure public security.[66]

The consideration of surveillance technologies in the Global South is timely. In Latin America, violence and security have become dominant themes over the last decade, following the processes of democratization from military and authoritative regimes during the 1990s.[67] While organized crime exists in the Global North, with gang violence affecting quality of life in neglected urban areas, its scale and intensity are dramatically higher in Latin America. Levels of delinquency have surged in many countries, especially Brazil, Mexico, Venezuela, and much of Central America.[68] Not only have governments been unable to deal with the violence, but organized crime now comprises the sociopolitical order.[69] States may use extralegal (or criminal) force against unions, gangs, or political opponents, while in other instances elements of the state may use organized crime to enrich themselves. Crime syndicates may buy off state actors in order to protect their operations, while in other instances violence is used as a means of conflict resolution where more formal legal channels are not available.[70] These "violent multiplicities"[71] highlight how the transition to democracy in Latin America has not brought a strong rule of law or civilian control of military forces, basic elements of democratic rule.[72]

Mexico is very much part of this story. The country has undergone a number of profound changes in recent decades. It experienced a progressive opening of its political system, highlighted in 2000 by the election of a president (Vicente Fox) from a party (the PAN, or National Action Party) other than the PRI (Institutional Revolutionary Party), which had dominated political life since the end of the Mexican Revolution in the late 1920s. The democratization of Mexico follows two decades of economic liberalization begun by President Miguel de la Madrid in the 1980s and reinforced with the North American Free Trade Agreement (NAFTA) in 1994, clear moves away from the protectionist, import-substitution economic model embraced by the PRI since the Great Depression. The political and economic transformations have been accompanied by cultural changes as well, as consumerism and globalization have reshaped national identity in the world's second most populous Catholic country. While the consequences of these processes are subject to debate, Mexico's entry into the twenty-first century promised a break from its more immediate past.

The massive violence of the last decade has cast a shadow over these developments, however. Much of the insecurity is of course driven by illicit drug trafficking contested by organized crime syndicates. Sharing a long terrestrial border with the United States, Mexico has been the preferred point of transit for cocaine shipments from South America since increased policing of the Caribbean Sea diminished the lucrativeness of water transits.[73] Since 1997, when Amado Carrillo Fuentes, leader of the Juárez cartel, died following plastic surgery intended to hide his identity, different cartels have been battling for control of the Juárez plaza (territory and supply route). This competition intensified in 2007 when Joaquín "El Chapo" (Shorty) Guzmán, leader of the Sinaloa cartel, allegedly was unable to come to an agreement with a rival cartel about access to Juárez.[74] The battles between the cartels have resulted in an increasing number of drug-related murders.[75]

However, drugs comprise just a portion of the crime problem in Mexico. While official crime statistics are unreliable, given the large amount of unreported crime *(cifra negra),* kidnappings, extortions, and street robberies have been on the rise.[76] Elevated crime also reflects the inefficacy of Mexican federal, state, and local governments to respond to delinquency. The country ranks high in international measures of impunity,[77] with national surveys indicating that over 93 percent of all crimes are either not reported or not investigated by authorities.[78] And half of the cases that are investigated do not result in further legal action.[79] As a result of insecurity and the state's inability to confront it, Mexicans report low levels of confidence in their police and public leaders.[80]

Upon assuming office in December 2006, President Felipe Calderón launched a War on Crime to counter this insecurity. Dropping past administrations' de facto policy of passivity and complicity toward the drug cartels, Calderón moved to disrupt their operations by reinventing public institutions and crime-fighting strategies.[81] National legislation was passed to increase punishments against criminals and to authorize the use of the military in fighting crime,[82] police forces were created and reorganized to reduce corruption and increase effectiveness,[83] international agreements were struck with neighboring countries to coordinate anticrime strategies,[84] national agreements were reached with the country's media to manage the content and tone of news reports,[85] and public relations campaigns were launched to increase public confidence and participation in crime reporting.[86]

Surveillance technologies have figured centrally in Mexico's crime fight. The Mérida Initiative signed by President Calderón and US president George W. Bush in October 2007 provided Mexico $1 billion for the purchase of advanced military and surveillance equipment.[87] Fusion centers have been established at different spots in the country to centralize and synthesize the analysis of crime data.[88] And a trio of innovative programs were set up during the Calderón *sexenio* (six-year term) with the intention of monitoring people and the things commonly used in the commission of crime: the National Registry of Mobile Telephone Users (RENAUT), a governmental database of cell phone lines and their subscribers, which would aid authorities in responding to kidnappings and extortion calls; the Citizen Identity Card (CEDI), a national identity card featuring biometric data such as fingerprints and iris scans, which would increase people's security from identity theft and fraud; and the Public Registry of Vehicles (REPUVE), a centralized federal registry of every vehicle in the country, along with attached RFID tags, which would combat car thefts, kidnappings, and drug trafficking.

The embrace of surveillance technologies to fight organized crime in Mexico invites a variety of questions. Have the surveillance technologies reduced insecurity? How have the technologies affected the government's ability to combat organized crime? How do people react to them? And what does the use of surveillance technologies tell us about the types of government we can expect in the future, both in Mexico and beyond?

1.4 SURVEILLANCE TECHNOLOGIES IN MEXICO'S WAR ON CRIME: A METHODOLOGICAL OVERVIEW

The pages that follow are an effort to answer these questions using a case-study methodology—studying particular instances of a phenomenon (in this case, governments' use of surveillance technologies against insecurity) in order to understand it in depth and in context.[89] I examine the three surveillance programs mentioned above: the RENAUT, CEDI, and REPUVE, with particular emphasis on the last program. I had originally planned to study the REPUVE exclusively, which appealed to my interest in cars, governance, and Latin America. Inclusion of the RENAUT and CEDI was motivated by what became a central theme for the book: the difficulties that authorities encounter in trying to implement monitoring programs. Arriving in Mexico in summer 2010 to begin preliminary research on the REPUVE, I was surprised to find, given the fanfare that had accompanied the program's launch in 2009, that few of my friends and acquaintances in Mexico City had heard of it.

Concerned that a fledgling program might fail to generate an adequate amount of data, I elected to research the other two programs as well.

To examine the programs, I relied on a triangulation of qualitative and quantitative methods and collected data over a four-year period, from 2010 to 2013. The first source of data were newspaper reports and government documents on the three programs that I collected primarily through the Google Alerts service for gathering Web content. The majority of this material came from national newspapers in Mexico, such as *El Universal, Milenio,* and *La Jornada,* and local papers from the areas where the programs were being implemented. I maintained a separate electronic file for each program. Altogether, these files comprised over three hundred thousand words. I used content analysis to identify key facts and themes.

A second source of data were observations I completed at two REPUVE registration sites in the states of Zacatecas and Sonora. During these field visits, I spent time with the staff responsible for inspecting vehicles and registering vehicular data in the REPUVE database and focused my attention on how the technologies worked in practice. These visits also enabled me to converse with staff and local program administrators to learn their opinions of the program and the successes and challenges they experienced. I also interviewed drivers who were registering their vehicles in order to understand their impressions of the REPUVE. These interviews usually lasted between five and fifteen minutes, and I interviewed thirty-five drivers in all.

I also interviewed eight national-level REPUVE administrators working at the offices of the Executive Secretariat of the National System of Public Security (SESNSP) in Mexico City. During separate visits to the SESNSP, I was able to speak with officials working in each of the REPUVE's four directorates: the State and Federal Operations Implementation Directorate charged with supervising the program's adoption by states *(entidades federativas)* and federal agencies *(autoridades federales);* the Relations with Obligated Subjects Directorate responsible for ensuring the compliance of private-sector businesses *(sujetos obligados);* the Procedures and Citizenry Directorate responsible for managing the public's contact with the program; and the Oversight and Verification Directorate charged with technical aspects of vehicle inspections. I usually conducted these interviews in small groups, organized by directorate, which facilitated scheduling.

The REPUVE directors at the SESNSP also arranged meetings for me with representatives from two automobile companies, which I refer to as Sucaro and the Veloz Motor Company (VMC).[b] I met with the Sucaro representatives at the SESNSP's Mexico City offices and with VMC staffers at the company's car plant. I also toured the VMC facility to better understand its process for complying with the REPUVE.

In Mexico City, I also visited ten auto dealerships to collect their impressions of the REPUVE program. These interviews, like the interviews with drivers, lasted between five and fifteen minutes. In addition, I attended a security technology trade show to gather observations and discuss the three programs with retailers. I followed this with an interview with a representative from a company that had unsuccessfully competed for the contract to supply RFID tags to the REPUVE. I conducted all interviews in Spanish, with the exception of that with the RFID company representative. Two undergraduate research assistants later transcribed the audio files, and I analyzed the transcripts by hand to identify key themes.

Finally, to better understand the public's reaction to the three programs, I created a sixty-item survey on the programs and insecurity in Mexico more generally. The survey provided respondents a description of each program culled from newspaper accounts and asked their opinions of them. Friends in Mexico helped me refine the instrument to improve readability. To gather data, I used snowball sampling with acquaintances in Mexico via Qualtrics and employed a data-collection company, Indagaciones y Soluciones Avanzadas, to distribute the survey in two sites that I could not access on my own—a working-class neighborhood in Mexico City and a rural community in Zacatecas. This purposive availability sample consisted of ninety-eight people. The data was coded into a SPSS file with the assistance of my research assistants.

These methods are not without their problems. An initial concern is the selection of Mexico as a site for studying surveillance technologies in the first place. Readers, especially Mexico scholars, might object that the country's unique historical and political development—for instance, the Spanish conquest and the lingering inequalities between European, mestizo, and indigenous populations; the Mexican Revolution's legacy of dislocating the country's elite and remaking the armed forces; and

---

b. Throughout, I use pseudonyms for the individuals and car companies I interviewed or observed. Providing research participants anonymity facilitated their participation and protected them from any possible reprisals their critical assessments of the programs or the government might risk. For individuals and companies mentioned in publicly available news reports, I used their true names.

the revolutionary state's corporatist framework and land redistribution program—limit its value as a case from which larger trends concerning surveillance technologies can be generalized. But a similar argument could be made about nearly any country—the peculiar longevity of the US Constitution, the history of slavery and racial inequality, the political and cultural legacy of the Civil War, the experience of westward expansion, and the unique geopolitical consequences of serving as a superpower have not prevented scholars from conducting research in the United States that informs social science more generally. What is more, as noted earlier, the insecurity and organized crime currently afflicting Mexico and the political strategies that authorities use to address it are very much regional and global trends.[90] Thus, rather than limiting ourselves from making larger statements on the basis of case-study methodology, I find it more useful to acknowledge the specificities of the settings studied, reflecting on how they affect the generalizability of a study's findings.

Beyond this, each of the methods utilized in this work possesses its own limitations. Media coverage is not the most reliable data source, given the political and financial concerns that inform coverage.[c,91,3] While I acknowledge this, my research relies on mainstream media sources that are generally viewed as reputable in Mexico, such as the newspapers *El Universal, Milenio,* and the current events weekly *Proceso.* In terms of both the observations and interviews, a main concern is the quantity of data. I focus on the REPUVE rather than the RENAUT and CEDI; my observations are limited to two field sites; and my interviews are with a select group of public administrators, workers, businesses, and users. I have little doubt that extending the interviews and observations to the RENAUT and CEDI would have yielded important insights on these programs and that conducting observations of the REPUVE in additional sites and interviews with more officials, businesses (such as insurance companies), and users would have produced distinct data. Nevertheless, the information collected does provide a robust view of the three programs, especially the challenges faced by the REPUVE. And the interviews produced common themes that, if not capturing the whole story of the REPUVE, represent key points for understanding its history.

c. To underscore this point for research on Mexico, in March 2011, representatives from various news outlets signed an accord with the federal government agreeing not to circulate stories that would threaten state operations against crime groups or cast the government in a negative light.

Finally, the survey instrument itself is limited because I did not distribute it to a random sample of Mexicans, it included a low number of respondents, and it lacked sophisticated measures of question reliability prior to circulation. Thus, the survey is not representative of Mexicans' thinking about the three programs and insecurity. But the survey results still provide rare insight into the thoughts of the public on the use of surveillance technologies in Mexico's War on Crime.

In sum, the methods used in this study possess not insignificant limitations that restrict the quantity and quality of the information examined. However, the variety of data examined here is a strength. And it proves adequate for discerning the histories of the three monitoring programs launched by the Calderón administration and reflecting on their significance for understanding the impact of surveillance technologies on contemporary governance more generally.

1.5 MAKING THINGS STICK: THE ARGUMENT IN BRIEF

This book's main argument is twofold. First, while surveillance technologies adopted in the name of security are generally understood as tools used by state authorities to monitor individuals, they are also used to monitor the things (automobiles, telephones, etc.) thought to underlie the commission of crime. It is not simply the individual driver, phone user, or name bearer that the REPUVE, RENAUT, and CEDI target, but the broader activities of automobility, mobile telephony, and personal identification in general. By adhering RFID tags to vehicles, having people register their phones, and creating identity cards based on biometric data, the state looks to gain purchase on the material basis of everyday life.

Monitoring mobility, communication, and identification are not new concerns for the state. Rather, they are inherent to "seeing like a state,"[92] strategies of governance that have been central to the formation of the state in Mexico over the course of its history. The Spaniards' military conquest of the Mixteca kingdom ruling the Valley of Mexico, the imposition of the Spanish language and naming practices on indigenous populations, the registration of individual citizens for the purposes of democratic elections following independence, the construction and securitization of roadways during colonial times, and the construction and securitization of railways during early modern times, among other examples, all represented efforts by authorities to control mobility, communication, and/or identification in order to realize social control. And such efforts were significant to the evolution of the state in Mexico.

But the creation of the REPUVE, RENAUT, and CEDI in the current day speaks to a crisis of governance—born of a society that has become increasingly difficult to administer over the past thirty years and a state apparatus that has become increasingly unable to govern society—that surveillance technologies are intended to address. In Mexico, law-enforcement officers and agencies are often corrupt; the data that the state generates on people and things are often inaccurate; and multiple agencies at local, state, and federal levels of government are often dedicated to the same task. Such obstacles within the state make governance challenging. Programs like the the REPUVE, RENAUT, and CEDI, by routing the information that RFID tags, mobile devices, and electronic identity cards produce through centralized databases, would enable the federal government to reduce its reliance on officers and agencies at the state and local levels that are seen to have become ineffective in the task of governance.[93] These programs would reform the state.

The surveillance technologies employed in the REPUVE, RENAUT, and CEDI possess, then, a dual purpose. They are intended to increase the federal government's grip over the mundane objects of everyday life, and they are intended to consolidate governmental authority over the administration of these things. I refer to this novel approach to governing as *prohesion,* a neologism formed from the Latin root, *haereo,* which means "to hang or hold fast, to hang, stick, cleave, cling, adhere, be fixed, sit fast, remain close." It is the root of the words "adhesion" and "cohesion," "adherence" and "coherence." And I use it in an allusive attempt to describe the efforts of authorities to make the materiality of social life more "adhesive" and the diverse organizations of government more "cohesive" to the state. To explain the title of the book, the surveillance technologies studied in this work are designed to "make things stick."

That is the first part of the argument. The second part is that the vision of state authorities in Mexico to "make things stick" has largely failed. Today, none of the three programs operates in the manner state planners had intended. To most Mexicans, this does not come as a surprise. While I was conducting research, many people I spoke with—whether they supported the programs or not—had a hunch that these programs would fail, like so many past efforts by the state.

But if failure is not surprising, one goal of this book is to try to bring the reasons for it to light. They are multiple. Ordinary people either refuse to register for programs or register using false information. Businesses push back against regulations that require them to add new procedures for complying with the law. Technologies fail to work in the manner expected. Resources are insufficient to successfully implement the programs on the scale desired. Political intrigue affects elected officials' willingness to follow the federal government's lead on security programs. States neglect to implement programs as they hold out for increased resources from the federal government. Thus, multiple points of resistance operate to weaken the federal government's efforts to remake the state through prohesion.

But weakness cannot be mistaken for absence. While the automobile registry, mobile phone registry, and personal identity card launched during the Calderón administration do not work in the way they were imagined, the programs, the ideas that inspired them, and the technologies that embody them operate in modified form. That they operate at all owes in good measure to the political acumen of program administrators to respond to the resistance the programs encounter and find points of connection with the individuals, groups, businesses, politicians, and things that oppose them. Programs are added onto existing state infrastructures in order to reduce costs, or they are reframed as intended for vulnerable populations in order to increase the programs' legitimacy, or they are eliminated altogether to allow the state to pursue alternative methods of managing mobility, communication, and identification. Such improvisational practices, which I refer to as *statecraft*, lend state formation an emergent quality, unknowable in advance but taking shape through practices and patterns of rule inherited from the past. If the fickle nature of statecraft proves discouraging for those in search of definite answers to how surveillance technologies will affect relations between governors and the governed in Mexico, it also ensures that a space for meaningful political action will continue to exist in our technological future.

1.6 CONTRIBUTIONS: SURVEILLANCE, STATE FORMATION, SOCIAL THEORY, AND LATIN AMERICA

In this work, the academic field I am in closest dialogue with is surveillance studies, and here, the points of connection are multiple. First, this work provides a more robust understanding of surveillance technologies by emphasizing dimensions of their operation beyond sight and vision. Our thinking on surveillance has been dominated by sight as a human sense. If cover art on leading social science books about surveillance is any measure, surveillance brings to mind CCTV cameras, computer monitors, magnifying glasses, X-ray images, and other devices designed to promote visibility. This emphasis prevails in the way surveillance gets discussed as well, as evidenced by such terms as "SuperVision,"[94] "stretched screens,"[95] or "visible war."[96] These images and this terminology communicate the idea that surveillance technologies increase the ability of authorities to watch over their subjects.

Besides being reasonable, this stress on visibility is important for understanding the shifting nature of power brought forth by surveillance technologies. CCTV cameras have become ubiquitous in the contemporary world, with Great Britain notoriously leading the charge with some six million devices for a population of sixty million people.[97] Cameras in mobile devices have increased imagery in the world, presenting new dilemmas for concerned parents of teens who share explicit images of themselves as well as new opportunities for "sousveillance," or the watching of authorities by those below them.[98]

But there are also reasons to be skeptical about the power of sight and our emphasis on it. Jean Baudrillard introduced the concepts of "simulacra" and "simulation" to comment critically on the changing relationship between image and reality in society. Today, we find ourselves moving "from a capitalist-productivist society to neo-capitalist cybernetic order that aims now at total control" through "the minute duplication of the real, preferably on the basis of another reproductive medium—advertising, photo, etc."[99] William Bogard has integrated these ideas into his work on surveillance, arguing that simulation "functions in ways that are totally contradictory to surveillance—not as a method of exposure or unconcealing, but the fabrication of completely original scenes, pure fictions that bear absolutely no relation to 'reality' at all, not even as a signified absence."[100]

Beyond the reality of digital imagery, a range of information technologies such as RFID tags, biometric cards, mobile devices, personal computers, and the networks that link these devices have transformed the nature of surveillance. Sensitivity to how the technical means of surveillance have transformed the nature of monitoring is captured well in the disparate definitions of surveillance offered by prominent scholars in the field.[d] Common to each is surveillance as the processing and analysis of data, or "dataveillance."[101]

With this broader field of surveillance studies in mind, the RENAUT, the CEDI, and the REPUVE in Mexico serve as detailed case studies of the technical and administrative operations required to erect a "surveillant assemblage" that collects data on telecommunications, personal identification, and automobility. These cases support the development of what can be thought of as a "material perspective" of surveillance technologies.[102] To create an electronic identity card based on biometric data, the human body is probed in different ways. Fingers are touched and recorded, irises are scanned and logged, and that data is encoded into barcodes and other formats stored in the card as well as government databases. To create an automobile registry, the body of the car is inspected and touched in order to record multiple instances of a vehicle identification number, and that data is then inscribed into government databases and RFID tags applied directly to vehicles' windshields. With these technologies, authorities seize directly on the body or materiality of the agencies of communications, identification, and mobility. This emphasis on touch and adhesion is why it is meaningful to speak of *prohesion*. If surveillance is understood as "watching over people" for the sake of affecting their behavior, the histories of surveillance technologies in Mexico reveal an operation in which authorities through technological means attempt to get and keep a hold upon the things that energize social life. Taken a step further, if the eighteenth and nineteenth centuries witnessed a "slackening of the hold of the body"[103] by authorities, who turned their focus to the soul, the contemporary world finds authorities renewing their interest in seizing upon the body.[104]

---

d. Gary T. Marx defines this new surveillance as "scrutiny through the use of technical means to extract or create personal or group data, whether from individuals or contexts," carefully choosing "the verb 'scrutinize' rather than 'observe' [to] call attention to the fact that contemporary forms often go beyond the visual image to involve sound, smell, motion, numbers, and words" (Marx, "Surveillance and Society," 2). David Lyon defines the new surveillance as "any collection or processing of personal data, whether identifiable or not, for the purposes of influencing or managing those whose data have been garnered" (Lyon, *Surveillance Society*, 2). Torin Monahan, for his part, studies "surveillance systems . . . that afford control of people through the identification, tracking, monitoring, or analysis of individuals, data, or systems" (Monahan, *Surveillance in the Time of Insecurity*, 8). And Kevin Haggerty and Richard Ericson describe a "surveillant assemblage" that "operates by abstracting human bodies from their territorial settings, and separating them into a series of discrete flows. . . . These flows are then reassembled in different locations as discrete and virtual 'data doubles'" (Haggerty and Ericson, "Surveillant Assemblage," 605).

A second point of engagement with surveillance studies concerns the state. Although policing, national security, and border control—activities where state authorities exercise legitimate force over populations—are common topics within surveillance studies, the state often only lies in the background. On the one hand, reflecting their debt to Michel Foucault and poststructuralist perspectives more generally, many studies view the adoption of surveillance technologies as indications of modes of thinking and acting—referred to variably as "governmentality,"[105] "risk,"[106] "technostalgia,"[107] and so forth—that operate behind the backs of authorities. On the other hand, some studies make reference to the state through the legislation (USA Patriot Act, US "Real ID" Act, Enhanced Border Security and Visa Entry Reform Act, etc.) and institutions (Department of Homeland Security, Transportation Security Administration, Federal Bureau of Investigation, etc.) that authorize and employ the surveillance technologies in policing and national security work.

These approaches are not without their strengths. The governmentality perspective identifies operations of power beyond particular institutional contexts and in seemingly benevolent actions undertaken by state authorities. And the legislation and institutions utilizing surveillance technologies are undoubtedly a central part of the story. I use these perspectives to describe the REPUVE, RENAUT, and CEDI in the following chapters.

However, these approaches also have their limitations. As other works have shown and the latter chapters of this book illustrate, what surveillance technologies do in practice vary from what they were designed to do.[108] And these divergences are very much tied to the structure and organization of the state and its relationship to society over time. State officials and federal agencies may have interests separate from national administrators charged with implementing monitoring programs. Inadequate resources can affect which elements of a program get implemented. And prior administrations' efforts to implement monitoring programs can make people mistrust current administrations' endeavors. By being sensitive to the complexity and dynamism of state forms and their relationship to society over time, this work looks to account for the role of the state in the outcomes of surveillance technologies.

With the previous point in mind, a third contribution of this work is to counter the dystopian normative stance of much of the surveillance studies literature. Often, scholars in the field hold the view that surveillance technologies will provide state authorities, and the private companies that work with them, increased control over people and will heighten stratification between different groups in society. In highlighting the struggles experienced by the RENAUT, REPUVE, and CEDI, this work challenges such assumptions. The federal government in Mexico consistently comes across as weak in this work, as companies oppose surveillance measures for the costs they add and people mobilize to block measures. The point is not that scholars should be cheerier about surveillance technologies. Rather, it is that the role of surveillance technologies in mediating our relationship with state authorities can only take shape through the messy operations of statecraft, making both negative and positive outcomes possible. As Gary T. Marx notes, "Perhaps then a nuanced perhapsicon model better captures our situation than either a panopticon or an utopicon model."[109]

Beyond surveillance studies, this work also speaks to research on state formation. Recent years have seen a number of excellent works inspired by science and technologies studies (STS) that illuminate how technoscience is implicated in state formation. The application of statistics, cartography, sanitary engineering, and the like in colonial Ireland transformed the island into a laboratory of English science and government and laid the foundation for the emergence of modern government.[110] Scientists and engineers, through agricultural, environmental, and military planning that helped exploit Saudi Arabia's vital nonpetroleum natural resources, gave shape to the basic institutions of that country's political system.[111] The Baku-Tbilisi-Ceyhan pipeline through Azerbaijan, Georgia, and Turkey is central to the construction of geopolitical order in the region and occasions new relations between nature, corporations, and publics.[112] And Stafford Beer's failure to cybernetize the socialist economy of Chile in the early 1970s was central to the inability of the Salvador Allende administration to build a lasting socialist state as an alternative to the international capitalist order.[113] Each in its own way, these examples illustrate that scientific knowledge and technological artifacts, in addition to coercion and capital,[114] are intimately involved in the "co-production"[115] of the state and social order.

This investigation into authorities' efforts to manage mobility, communication, and personal identification over the course of Mexican history contributes to this trend. The colonial government's attempts to secure roadways buttressing the extraction of silver and other natural resources gave birth to the Tribuna de la Acordada, a policing organization charged with protecting roadways from bandits. Efforts to implement voting procedures in the country following independence from Spain led to the creation of both mailed ballots—de facto identity cards for individual voters—and governmental agencies to administer them. And interest in bolstering support for the postrevolutionary state in Mexico led authorities to construct schools throughout rural Mexico that would combat the hold of the Catholic Church on the Mexican worldview. Material artifacts, in brief, have served to co-produce the Mexican state.

But this study also provides deeper insight into the processes of state formation. Older forms of state organization and practice that first emerged to manage social activities such as interpersonal communication—a state telephone monopoly, for instance—can lose their grip on society as new technologies (e.g., the mobile phone) transform the practice of communication. This leads state authorities to consider new tactics—the surveillance technologies of the RENAUT, the CEDI, and the REPUVE—to regain their grip on the social order. But older forms of government do not simply wither away as their utility diminishes. Aging institutions, bureaucracies, offices, actors, and patterns of behavior present a counterweight to new ways of governing. Thus, state forms and practices must continually be negotiated by administrators through the practice of statecraft, which gives the state a dynamic, indeterminate quality. The state is continually being remade, built on infrastructures and patterns of interaction inherited from the past.

This emphasis on the evolutionary indeterminacy of state forms contributes to theorizing on social order more generally. In recent years, the social sciences have shown heightened interest in incorporating what has been referred to as posthumanist thought from STS as a way of moving beyond the discursive focus of poststructuralism and social constructivism. Posthumanist thinking, or "assemblage theory,"[116] emphasizes the materiality of the social world, the distribution of agency between human and nonhuman actors, and the emergent nature of social order.[117]

This theoretical current informs the present work in multiple ways. For instance, the primary object of the surveillance programs studied here are the material things (automobiles, mobile devices, and body) through which distributions of agency (mobility, communication, and identification) circulate in society. *Prohesion* is designed to make these things stick. The durability of social assemblages is vital as well, as past forms of governance can resist the efforts of government leaders to adopt new ways of administering the world. And, as noted above, the form of the state emerges in time through the *statecraft* of administrators and bureaucrats in countering the resistance to their efforts.

But this work also cautions against overemphasizing the nonhuman in the emergence of social order. Studies adopting the language of STS and assemblage theory often approach social formations as self-organizing systems, a view perhaps nourished by the fact that the natural world has often served as a muse in key texts in the field.[118] But if agency in the world is distributed between humans and nonhumans, this invariably means that cultural beliefs, skills, and practices are still central to social outcomes. In the case of the REPUVE, RENAUT, and CEDI, it is the program directors and administrators working behind the scenes who help determine which pieces of old programs and infrastructures can be repurposed for integration with new ones, which businesses can be persuaded to be early adopters of new programs, which state actors can be negotiated with and convinced to implement them, which technical problems can be fixed and which cannot, and so forth. The word "statecraft," like "bricolage"[119] and "metis,"[120] is intended to highlight this practical aspect of social order. And these examples encourage us to remain attentive to the potential of human action to influence and remake the world.

Finally, this work is also in dialogue with research on politics and insecurity in Latin America. As noted earlier, following a period of democratization in the 1980s and 1990s, the liberalization of the region's economies, the accompanying retreat and contraction of the state, and the integration with the global economic order have provided fertile ground for the intensification of illicit drug and human trafficking by organized crime syndicates.[121] These criminal organizations have penetrated the state to become integrated elements of the social order.[122] Security has as a result becomes a key policy focus for governments in the region, defined by an increasing militarization of police forces, the adoption of legal reforms that expand the coercive power of the state at the expense of civil liberties and protections, and regional collaborations with neighboring governments to coordinate crime-fighting efforts.[123]

The present work adds to that literature by emphasizing the role of technology in both the creation and response to insecurity. The growing use of mobile devices, automobiles, and online identities by large numbers of people in Mexico presents new challenges to state authorities looking to combat crime amid institutional corruption and in a neoliberal era of reduced budgets. And surveillance technologies such as RFID tags, GPS devices, biometrics, and computer networks are embraced as a way to reshape the state to meet the security challenges of the day. Thus, to better understand security measures in contemporary Latin America, we must squarely focus attention on the technologies and material things of modern life and governance.

In addition, to the extent that we scholars who study Latin America are invested in social outcomes in the region, this work provides some measure of hope. While some studies exhibit a measure of skepticism or uneasiness about Latin American governments' adoption of surveillance technologies,[124] stances that simulate the dystopian trends of surveillance scholarship more generally, this work illustrates how surveillance technologies can bring about a remaking of the state that requires the participation of ordinary citizens. In this sense, surveillance technologies can serve as an opportunity for people to engage with the state and, in the process, strengthen democracy and the rule of law.

1.7 AN OVERVIEW OF THINGS TO COME

Over the next five chapters, I develop these ideas in greater depth. Chapter 2, "Taming the Tiger," begins with a description of the Registry of Mobile Telephone Users, Citizen Identity Card, and Public Registry of Vehicles and then considers them in light of larger historical trends in Mexican history. It argues that while the RENAUT, the CEDI, and the REPUVE resemble programs described by surveillance scholars elsewhere, they are most noteworthy for their focus on the materiality of communication, identification, and mobility. Authorities' efforts to administer these activities are not new; rather they date back to the Spanish conquest and have over the course of Mexican history "co-produced" the state. The launching of the RENAUT, CEDI, and REPUVE in Mexico today, however, speaks to a crisis of governance that the democratization of the country's political system, liberalization of its economy, and transnationalization of its culture—all aspects of globalization—have brought about.

Chapter 3, "Prohesion," delves into the operation of surveillance technologies by locating the REPUVE within the history of automotive governance in Mexico. This chapter argues that the REPUVE evidences a third historical approach to governing the insecurity of automobility. Discipline was first pursued to increase the safety of road travel by making drivers responsible. Risk management later emerged to reduce the environmental harm of automobility by monitoring car emissions.

The third phase, which I term "prohesion," operates not by targeting the subjectivity of drivers or the invisible emissions of vehicles, but by gaining a hold of the materiality of vehicles in order to achieve "legal certainty" about automobiles. The bodies of vehicles are inspected and registered into the registry, while RFID tags are adhered to their windshields in order to provide the state a constant presence or hold on vehicles. Through prohesion, the federal government in Mexico attempts to respond to the challenges of globalization by making the material things of society and the administrative agencies of the state stickier and more cohesive.

Chapters 4 and 5, *"Ni Con Goma"* and "Statecraft," examine the difficulties inherent in trying to make things stick. Chapter 4 reviews the different points of resistance encountered by the REPUVE, RENAUT, and CEDI programs in Mexico. As noted above, ordinary people, corporate actors, technological artifacts, and state entities all in their own way defy the federal government's efforts to gain a grip over mobile telephony, personal identification, and automobility. Thus, resistance is inherent in prohesion and threatens state designs to control these collective activities. But more than this, opposition to the programs demonstrates the force of history in Mexico, where particular formations or "assemblages"[125] of personal identification, mobility, and communication and their governance forged during earlier periods persist in the face of official efforts to reshape them. Taking these lessons to heart, the chapter seeks to broaden our conception of resistance in sociopolitical contexts to explain the weakness of surveillance technologies as tools of governance.

In response to such resistance, authorities in Mexico have needed to make various alterations to the RENAUT, CEDI, and REPUVE. To reduce popular resistance, administrators tie theses initiative to existing programs and institutions that ordinary Mexicans already deem legitimate. To overcome private-sector resistance, the federal government does not penalize companies that at least show good faith in implementing the programs. To overcome state resistance, federal authorities allow local government to charge people for enrolling in the registries, seemingly in violation of the law. In brief, *prohesion,* or making agencies stick, requires authorities to engage in *statecraft,* modifying program requirements and operations on the fly. Administrators who are able to put surveillance programs into action in Mexico through statecraft alter the programs' function and meaning, giving the state an emergent quality. While such accommodations might seem to evidence the chaotic nature of the Mexican state, they are better considered as signs of a collaborative impulse that bode well for Mexico's democratic future.

As these summaries suggest, this work offers a different conception of surveillance technologies than that generally encountered, one that is wider in scope and hopefully better suited to comprehending the workings of governance in global society. In the Chapter 6, "Grasping Surveillance," I reflect on the theoretical lessons of this study in terms of four themes: tactility, weakness, emergence, and engagement. These are meant as conceptual counterweights to tendencies that often frame our thinking on surveillance: visibility, strength, determinism, and fatalism. While the tactility of prohesion as a security strategy provides an ominous vision of a future that surveillance technologies could usher in, these technologies' weakness provides the solace of knowing that political engagement will remain central to whatever outcomes emerge. With these lessons in mind, the work finishes by considering what courses of action subjects in the surveillance society might meaningfully take to help create more democratic futures.

Taming the Tiger

Madero has unleashed the tiger, let's see if he can tame it.
—Porfirio Díaz

2.1 SURVEILLANCE TECHNOLOGIES AND MEXICO'S WAR ON CRIME

Mexico's first Citizen Identity Card (CEDI) was issued on January 24, 2011, to Leslie García Rodríguez, an eleven-year-old student at the Miguel Hidalgo Elementary School in Tijuana, Mexico. To mark the event, an energetic crowd assembled outside the school—named for the revolutionary priest whose call for independence from Spain two hundred years earlier had given birth to the Mexican independence movement. One by one, students, teachers, civic activists, and public officials took to the microphone to extoll the virtues of the identity card. By housing young people's biometric identifiers—iris scans, fingerprints of all ten fingers, and a photograph—in a single card, the CEDI would "guarantee the identity" of each young person in the country and serve as a vital tool against the trafficking of minors. The city mayor noted the symbolism of launching the program in Tijuana, a major migration conduit to the United States especially susceptible to such trafficking. Secretary of the Interior Francisco Blake, the highest-ranking federal official at the event, explained that with the card, "you can be assured that the right of each of you to an identity, a name, a family, and a country is fully guaranteed."[1]

Surveillance technologies have been growing steadily in Mexico, paralleling their spread across the world. Market dynamics here, as elsewhere, have helped drive the expansion. Private firms specializing in security services have exploded on the national scene. In 1989, 210 companies were registered in the area of protection; by 2004, there were 2,126. These firms generally specialize in three security sectors: information security; access to shops, homes, and factories; and vehicle reinforcement.[2] But if the private sector has been driving growth in security technologies, the public sector has laid the groundwork. Various municipalities have constructed "command centers" *(centros de mando)* that provide surveillance of city locations through closed-circuit television (CCTV) cameras.[3] Often installed to facilitate the flow of traffic, these technologies lend themselves to security operations because both emergency-response workers and the police find them essential to their work.[4] Police will force their way onto traffic-monitoring systems in order to track not only criminal suspects but also street protesters, while traffic operators use the cameras to detect and report pickpockets.

As noted in the introductory chapter, surveillance technologies have featured centrally in the War on Crime launched by President Felipe Calderón. The Mérida Initiative, signed by Calderón and US president George W. Bush in October 2007 provided $1 billion for the purchase of eight Bell transport helicopters, two surveillance planes, inspection equipment such as X-ray scanners, and modernized communications and information technology.[5] The Obama administration later extended the funding, set to expire in 2010, through 2012 at some $300 million per year.[6] Reflecting the increasing security cooperation between the two countries, US drones now fly over Mexico to collect data on drug cartel activities,[7] while Mexican drones fly (and crash) over the United States in similar intelligence-gathering operations.[8] The two countries have also established at least two fusion centers in Mexico that allow for the integration and analysis of diverse types of data related to drug trafficking.[9] It was as part of this push to combat organized crime that President Calderón launched the Citizen Identity Card (CEDI), National Registry of Mobile Telephone Users (RENAUT), and Public Registry of Vehicles (REPUVE).

These three programs serve as this work's case studies, and this chapter connects them both to prevailing theories of surveillance and broader trends in Mexican history. First, while the RENAUT, CEDI, and REPUVE bear a clear resemblance to operations described by surveillance scholars elsewhere, these programs are most noteworthy for their focus on communication, identification, and mobility, collective activities central to the operation of society. Thus, rather than simply targeting individuals, the programs also attempt to monitor the things—phones, bodies, and cars—through which agency is exercised in society. Second, while the technologies involved in the three programs are novel, authorities' concern with controlling these dimensions of social life is not. Rather, such efforts date back to the Spanish conquest of Mexico and have served over the course of Mexican history to help "co-produce"[10] the state. Third, if authorities' efforts to administer communications, identification, and mobility have helped form the Mexican state, the appearance of the RENAUT, CEDI, and REPUVE today speaks to a crisis of governance in the country, whereby the "vibrancy"[11] of society has outgrown the state's ability to manage it, resulting in increased levels of insecurity. These programs featuring surveillance technologies thus represent a concerted attempt to remake the state.

2.2 SURVEILLANCE TECHNOLOGIES, AGENCY, AND MATERIALITY

Of the three programs examined, the National Registry of Mobile Telephone Users was the first to come to the public's attention. Created in August 2008 through an omnibus security law, the National Agreement on Security,[12] the RENAUT called for the creation of a governmental database of cell phone lines and their subscribers and required cell service providers to maintain their own databases to store subscribers' names, addresses, fingerprints, and photographs and to provide geolocalization of individual calls, all of which would aid authorities in responding to kidnappings and extortions.[13] As well, the program demanded that service providers cancel service for phones reported stolen, thereby denying kidnappers the tools—stolen phones—regularly used in their crimes.[14]

The idea for a national cell phone registry emerged following a pair of high-profile kidnapping-murders. In September 2007, nineteen-year-old Silvia Vargas, daughter of Nelson Vargas, former director of the National Sports Commission, was seized and, following failed negotiations, found dead over a year later in the basement of a Mexico City home.[15] In June 2008, fourteen-year-old Fernando Martí, son of Alejandro Martí, a sports-apparel mogul, was kidnapped while being driven to school and later found dead in the trunk of an abandoned car after two months of confused negotiations.[16] Alejandro Martí subsequently founded the civic group Mexico SOS (Public Security Observation

System), which has played a prominent role in critiquing government security efforts and has worked to reform justice administration in Mexico. Increased vigilance over cell phones was an idea promoted by the organization from its founding.[17]

To implement the program, the Secretariat of the Interior (SEGOB) required cell users to register their phones with the RENAUT. Users could register either by sending a text message with their names and date of birth or Unique Population Registry Code (CURP), Mexico's equivalent of a national identification number, or by going to a service-provider center to have their data entered.[18] In February 2009, the government announced that cell users would have until April 10, 2010, to register their phones or have their service cut.[19]

The Citizen Identity Card, meanwhile, was announced in July 2009. The CEDI is a 3.4-by-2.1-inch plastic card containing the following information: two photographs (one embossed using the person's CURP or national identification number), name, birth date, bar code containing the person's CURP, and double bar code containing the person's iris scans.[20]

The card, supporters argued, would hold multiple benefits for Mexicans. First, it would facilitate daily transactions by providing a single form of identification to replace the innumerable forms that Mexicans currently possess—birth certificate, driver's license, voter card, advanced electronic signature for tax payments, military service card, and so on. Second, it would increase people's security from identity theft and fraud by providing a more sophisticated technological basis for establishing identity.[21]

The legal basis for the identity card derives from reforms made in 1990 and 1992 to the country's Population Law, in which the Salinas administration, in the context of rising migration to the United States, called for the creation of a National Population Registry (RENAPO) and the distribution of Citizen Identity Cards for those registered.[22] Similar to the RENAUT, participation in the CEDI program was obligatory. Mexicans would attend service centers where an array of technological devices provided by the Smartmatic Corporation[23]—computer laptop with Ethernet interface, fingerprint capture device, signature capture device, iris capture device, document scanner, bar code reader, among other things—would record biometric measures and legal identity. The cards themselves would be distributed later.

The third program—the Public Registry of Vehicles—received much less attention in the media. The REPUVE possessed three objectives: (1) to create a centralized federal registry of all cars circulating in the country, including vehicle identification number, registration information, physical description, and the name and address of owners; (2) to attach 18000–6C radio-frequency identification (RFID) tags to vehicles containing the unit's registration details; and (3) to install RFID readers and license plate recognition cameras at tollbooths and other transit points to verify the status of passing vehicles.[24] First announced on March 3, 2008, by the head of Mexico's Secretariat of Public Security, Genaro García Luna, the technology utilized in the registry was to serve as a tool to combat crimes involving automobiles, including car thefts, kidnappings, and drug trafficking.[25] Enrollment, as in the cases of the RENAUT and CEDI, was mandatory.[26]

President Calderón placed the first RFID sticker on the inside windshield of a Chevrolet Suburban at the Puente de Ixtla tollbooth on the Cuernavaca-Acapulco Highway outside Mexico City in June 2009.[27] In the following months, several states began installing tags on their public vehicle fleets. And a few states, Zacatecas and South Baja California among them, soon extended the provision to private vehicles as well.[28]

At first blush, the RENAUT, CEDI, and REPUVE appear to share much in common with surveillance technologies described elsewhere. The programs involve a variety of technological devices: identification technologies such as documents, numbers, RFID stickers, and biometric registers that affix identities to people, phones, and vehicles; information technologies in the form of integrated databases that store and share these identifying data along with geographic location; imaging technologies—video cameras—that read license plates and verify identities; and monitoring technologies, including RFID readers and phone logs, that track people and their things. The function of this technological assemblage is to monitor. These are surveillance systems, then, but with a high level of sophistication. They are not mere tools—such as thermal-imaging technologies[29] or crime-analysis software[30]—that security agents deploy to see things that otherwise would not be seen. Rather, they are examples of the "new surveillance"[31] or "dataveillance"[32] that, through the extraction and management of data about people and things, allows for constant, automated monitoring and identification across multiple physical, analog, and digital spaces.

The goal of this surveillance is to recognize and disable phones, cars, and people that carry suspect identities. In this respect, the programs share an affinity with surveillance projects described elsewhere, such as the NEXUS[33] and US VISIT[34] programs at the US border that seek to distinguish suspicious people who would harm the country from trusted global or "neoliberal citizens" who would contribute to the country's prosperity. The governmentality at work in the programs is thus similar as well. Matching Michel Foucault's description of the "double system" at work in "security," where the state deploys "great mechanisms of incentive-regulation" to ensure "the security of the natural phenomena of economic processes or processes intrinsic to population" together with "simply negative functions" to suppresses elements of disorder and illegality, these programs aim to separate out trustworthy callers from those who would use phones to extort money, trustworthy car sellers and buyers from those who would deal in stolen vehicles, and trustworthy people from those who would nab others or their identities for illicit ends.[35] This is surveillance for the sake of security, then, which aspires to the "social sort"[36] of citizens, phones, and cars to create a vibrant social environment secure from the specter of criminal activity. In sum, the adoption of surveillance technologies in Mexico could be read as part of a more general global shift to security governance, whereby public and private authorities adopt advanced technologies in order to sort flows of people and things and better control society.[37]

While such an interpretation is not wrong, a simple application of the surveillance literature, which is largely based on empirical research in the United States, Canada, and Great Britain, to the Mexican context risks overlooking other dimensions of the technologies central to understanding their adoption in Mexico and more generally.[38] As a first step, for instance, we can observe that the RENAUT, CEDI, and REPUVE target fundamental aspects of social life tied to insecurity and organized crime. The national cell phone registry responds to the fact that government possesses no reliable way to track the *communications* of kidnappers, who use either victims' phones or units purchased on the black market to negotiate ransoms. The national identity card addresses the considerable difficulties the Mexican state has encountered in *identifying persons*. In early 2013, it was reported that some fifteen thousand corpses had been encountered during the Calderón sexenio that were never identified before being buried in common graves.[39] The car registry, meanwhile, is meant to tackle the *mobility* of crime, both the high number of auto thefts[40] and the central role of automobility in the kidnappings and drug trafficking. In seeking to increase the government's control over mobile communication, personal identification, and automobility, the programs can be read as attempts to order the collective agency of society, or those practices that make collective life possible.

The concern with monitoring communication, identifying persons, and tracking mobility is not specific to the modern Mexican state. As James Scott explains in his seminal book, *Seeing Like a State,* states have always sought to control these human capacities. Scott writes, "Nomads and pastoralists (such as Berbers and Bedouins), hunter-gatherers, gypsies, vagrants, homeless people, itinerants, run-away slaves, and serfs," people whose mobility and means of subsistence make it difficult to tax, "have always been a thorn in the side of states."[41] As a result, "the precolonial state was . . . vitally interested in the sedentarization of its population—in the creation of permanent, fixed settlements."[42] The ability to order populations, in turn, rests on "state simplification," schemes for making people's daily lives legible to the state and open to intercession. Of these schemes, the most central perhaps is "the imposition of a single, official language," which standardizes modes of communication in a given territory.[43] Critical, too, are the methods by which people identify themselves. "The assigning of patronyms by family," Scott observes, "was integral to state policy promoting the status of (male) family heads, giving them legal jurisdiction over their wives, children and juniors and, not incidentally, holding them accountable for the fiscal obligations of the entire family."[44]

Connected to these efforts is a concern with the materiality of collective agency. Communication in contemporary society is increasingly predicated on "mobile telephony,"[45] which requires mobile devices linked by radio frequency to a larger technological infrastructure of cell phone towers, telephone exchanges, and fiber cables. Similarly, movement in modern society is based disproportionately on "automobility,"[46] which requires less the capacity to move oneself than the dexterity to operate a complexly simple machine that will do so in one's place. Finally, identification is carried out not so much by naming practices as through numbers and biometric data inscribed on "national identity cards" and documents that, like a common, national language in prior ages, can meet the "need for universally acceptable tokens of identification."[47]

These examples illustrate the point that people's ability to partake in the tasks of collective life—communication, mobility, and self-identification—is mediated in nearly every instance by technology. This understanding is a point of departure for science and technology studies (STS) scholars who have produced a robust collection of work evidencing how society consists of "actor networks" involving both human and nonhuman actors.[48] Automobility requires both the driver and car, not to mention the enormous assemblage of people and things needed to train drivers, manufacture vehicles, and keep both in functioning order. In short, the capacity to communicate, move, or identify oneself is no longer inherent or exclusive to the individual person. Rather, agency in society is "distributed" across an architecture of material things that gives agency to those connected to it.[49]

In their focus on cell phones, motor vehicles, and human biology, the RENAUT, REPUVE, and CEDI, respectively, reveal the sensitivity of governmental authorities to the importance of materiality within collective agency. In order to get a grasp on communications in contemporary society needed to combat the kidnapping of people, in order to take hold of mobility to stop the transportation of illicit drugs and weapons, in order to account for the identification of people to investigate crime, the RENAUT, REPUVE, and CEDI target the materiality of these activities.

2.3 COLLECTIVE AGENCY AND THE CO-PRODUCTION OF THE STATE ACROSS MEXICAN HISTORY

While the significance of materiality to projects of social ordering has only more recently become apparent in the social sciences, political authorities have long been aware of it. Indeed, as this section emphasizes, efforts to control communication, identification, and mobility can be found throughout Mexico's history. And these efforts to capture and control agency in society have played a key role in the "co-production" of the Mexican state.

2.3.1 *The Conquest of the Mexica*

It is accepted today that the Spanish conquest of the Aztec empire (more accurately called the Triple Alliance of the Mexica, Texcoco, and Tlacopan peoples, or the Mexica empire, since the Mexica held the most influence in the alliance) by Hernán Cortés and his expedition of six hundred men owed primarily to the power of smallpox, which the Spanish carried with them from the European continent, and political alliances, which Cortés struck with rivals, such as the Tlaxcalteca, whom the Triple Alliance had subjugated. Without the force of disease and the vast numbers of warriors offered to the Spaniards, it is improbable that Cortés would have been able to fell the mighty Mexica empire.

But underpinning this history as well was the expedition's ability to penetrate the Mexican system of communication and mobility.

Communication was particularly vital to the Spanish cause. The Spanish had no knowledge of the major language groups spoken in Mesoamerica—Nahuatl in the Valley of Mexico and Mayan on the Yucatán Peninsula. And this would have hampered their ability to negotiate alliances to conquer the Mexica had it not been for their chance encounter with two extraordinary historical figures—La Malinche and Gerónimo de Aguilar.

La Malinche was an indigenous woman born into a noble family but later given away to a community in Tabasco once her father died and her mother remarried. The unfortunate circumstance left her knowledgeable of both Mayan and Nahuatl dialects. When the Cortés expedition defeated the Tabascans, the Tabascan *cacique,* or chief, offered Cortés twenty slave women as a peace offering, a group that included La Malinche. Gerónimo de Aguilar, meanwhile, was a Spaniard who was part of an earlier expedition of the Yucatán Peninsula that had shipwrecked. Most on board were enslaved by a local indigenous group and perished either through hard labor or human sacrifice. De Aguilar escaped his captors and took refuge with another community of natives inhabiting lands near the island of Cozumel. Through his ordeal, de Aguilar learned the local Mayan dialect and thus could communicate between the Spanish and Mayans. When he learned of the arrival of the Spanish at Cozumel, de Aguilar and a group of natives raced out in canoes to meet them and the Spaniard joined Cortés's company.[50] Together, Doña Marina, the name under which La Malinche was baptized, and Gerónimo de Aguilar enabled Cortés to communicate and negotiate with the native groups he encountered on the way to Tenochtitlan, the capital of the Mexica empire.

The importance of the translators was lost on neither the Spanish nor the peoples they came into contact with. Doña Marina formed such a vital part of the Spanish company that the Mexica actually referred to Cortés as Malinche, revealing how they saw Cortés and Marina as a single, unified being (fig. 1). Bernal Díaz del Castillo, chronicler of the campaign, held no reservations in assessing La Malinche's contribution. "This woman was a valuable instrument to us in the conquest of New Spain. It was, through her only, under the protection of the Almighty, that many things were accomplished by us," he wrote. "Without her we never should have understood the Mexican language, and, upon the whole, have been unable to surmount many difficulties."[51]

The Spaniards' ability to overcome the mobility of the Mexica was also central to the conquest. In obvious ways, the Spanish, with

FIGURE 1. Hernando Cortés, 1485–1547, Spanish conquistador, with the Indian interpreter Marina, or the Malinche, and Aztecs. From copy of 1632 Codex Tlaxcala, National History Museum, Mexico City. Photo by Gianni Dagli Orti/The Art Archive at Art Resource, NY.

their massive sea vessels armed with cannons and horses that could be mounted for combat, enjoyed clear advantages in terms of mobility. That said, those central components of European warcraft proved of marginal utility to Spanish attempts to conquer Tenochtitlan. Constructed on an island in Lake Texcoco, the Mexica capital was adjoined to land by three major causeways. The Spanish were famously awestruck upon first seeing the city. "When we gazed upon all this splendour at once, we scarcely knew what to think," Bernal Díaz remembered, "and we doubted whether all that we beheld was real. A series of large towns stretched themselves along the banks of the lake, out of which still larger ones rose magnificently above the waters. Innumerable crowds of canoes were plying everywhere around us; at regular distances we continually passed over new bridges, and before us lay the great city of Mexico in all its splendor."[52] The aquatic location could be defended by thousands of warriors patrolling the lake with canoes. While the Tlaxcalteca and other groups allied with the Spanish had their own canoes, the numbers were unfavorable. The physical layout of the city thus rendered useless the Spaniards' ostensible advantage in mobility. And during Cortés's first advance on the city, the vast numbers of Mexica and allied warriors were able to stymie the Spanish.

In an attempt to change the balance of power, Cortés decided to lay siege to the city by destroying the aqueducts that brought fresh water, negotiating alliances with the other peoples residing near the lake and mounting a blockade aimed at depriving the city of provisions. To carry out the siege, Cortés ordered the construction of twelve brigantines, small fortified ships. The brigantines were constructed in Tlaxcala and outfitted with gear and weaponry from the Spanish fleet, which Cortés had disassembled to disabuse his company of any thought of abandoning the mission. Once launched on Lake Texcoco, the ships enabled Cortés to patrol the waters to prevent food and water from arriving and giving cover to Spanish and Tlaxcalan troops approaching on Tenochtitlan's causeways. With the mobility of the Mexica neutralized, Cortés was able to advance on and eventually seize their capital.

2.3.2 *Ordering Colonial Life in New Spain*

The Spanish conquest of Mexico was thus based in large measure on the ability of the invaders to seize the communications and mobility that made the Mexica and Triple Alliance dominant in the Valley of Mexico. Generally speaking, the conquistadors did not concern themselves with modes of personal identification among the peoples they conquered.[a] This began to change, however, with the establishment of colonial order in the land the Spanish coined New Spain.

a. The Catholic Church was in a different position, however, its function in the conquest the conversion of souls. The clergy did stand alongside conquistadors when subjugating native peoples, reading the Requirement that announced Spain's divine right to conquest for the purposes of evangelization. But the salvation of souls, as per Christian tradition, also required baptism. And to complete and keep records of baptisms, the church assigned individual identities to the native population. Individuals were usually given two names, both first names, an accommodation to the fact that last names were not a convention in pre-Hispanic Mexico. The first name was almost exclusively Spanish, following the most popular saints. Second names were more varied: Spanish names (Juan Diego), Nahuatl names (Juana Quautzontecontzin), and other designations announcing social position (Doña María Xalatlauhco) or devotion (de la Cruz, "of the Cross"). Nahuatl men usually possessed more distinction in their second names than women, which reflected that they could hold local political office. Names thus not only served the administrative purposes of the church but also inscribed the unequal social order enforced by the Spanish (Horn, "Gender and Social Identity," 119–22).

The political economy of the Spanish empire was based on the extraction of precious metals from its colonial holdings. Internally, the colonies required agriculture and livestock to subsist. Both activities required a sizable workforce. And in New Spain, the conquered peoples would presumably provide this labor.

In subjugating the native groups of New Spain within the colonial order, the Spanish relied on existing structures of indigenous social hierarchy. The conquistadors, initially with the blessing of the Crown, established the *encomienda* system, a feudal structure under which vast tracts of land and the peoples living on them came into the possession of individual members of Cortés's expedition. Indigenous villages were ruled by local chiefs, *caciques,* who were responsible for collecting tribute from their subjects, which included forced labor in the mines.[53] In political terms then, "things did not change as much as one would have thought" for the ordinary *indio* in Mexico, as systems of tribute and hierarchies remained the same, save for compulsory military service and human sacrifices.[54]

The boundaries between these two societies quickly broke down, however. The paucity of Spanish women in New Spain resulted in unions between Spanish men and indigenous women, resulting in *mestizo* children.[55] Disease and hard labor also decimated the indigenous populations,[56] which led to a labor shortage and the introduction of African slaves to the colony.[57] The eventual and ongoing mixing of people "produced a tertiary, intermediate people identified as castas" in Mexico.[58] Already in the 1540s, the ruling elite in New Spain identified *castas* as a threat to the social order, "the colony's foremost partisans of insurrection."[59]

In response to the perceived menace of mixed races, colonial authorities developed a caste system by which they could identify and order the evolving society. At the core of the *casta* system were five basic races: *españoles, indios, mestizos, negros,* and *mulatos.*[60] From this emerged different combinations—*zambo,* for instance, which designated people who were half Indian and half African.[61] Over time, more than forty categories were created, even if most lacked practical significance.[62] *Castas* varied in terms of rights and responsibilities. *Castas* in the 1500s were forbidden from living in indigenous villages or neighborhoods in town: "The crown in effect assigned the castas to the Spanish community."[63] *Negros* and *mulatos,* like *indios,* had to pay tribute. However, they could not dress like *indios* or in gold jewelry, indicative of European ancestry. Nor could they hold office or bear arms. *Mestizos,* meanwhile, did not have to pay tribute. And unlike *indios,* who were considered *gente sin razón* (people without reason), essentially children who needed to be cultivated into Christian civilization, *mestizos* were *gente de razón* (people with reason).[64,b]

With such measures, colonial authorities sought to identify and order people in New Spain through an examination of the body. This anchoring of personal identity to the body was reinforced through technological means as well. Separate marriage registers existed for different *castas.*[65] And *pinturas de castas* become a popular form of cultural expression during the eighteenth century (fig. 2).[66] The paintings, which depict the different race combinations, possess a clear pedagogical function in explaining the *castas* and their hierarchy to the viewer. And in doing so they served to reify the largely arbitrary racial classifications that authorities sought to impose upon the people of New Spain in order to expand their political control.

While the entrenchment of Spanish rule in Mesoamerica involved controlling the material aspects of communication, such as destroying indigenous religious artifacts[c] and establishing schools[d] to train children of the indigenous noble class, the Spanish Crown's management of mobility is more relevant to the current discussion. Gold most appealed to the Spanish, but lands to the north in and around the city of Zacatecas proved richer in silver deposits. And mines were established throughout the region. The transfer of wealth back to Spain, meanwhile, required a system of roadways that joined the mines to the colonial capital and the main seaport of Veracruz. Thus was born the Camino Real de Tierra Adentro, the Royal Road of the Interior Lands, which eventually

b. Race thus functioned differently in New Spain than in North America. If in North America "a single drop of black blood sufficed in staining people, in the Indies, a single drop of white blood precipitated a 'whitening' of people." For this reason, *indios* might try to escape the obligations of tribute and the political and resource limitations of village life by entering the labor markets and racial castes of the *mestizaje* (García Martinez, "Los años de expansión," 220).

c. Interestingly, Juan de Zumarrago, the first bishop of New Spain who was also given the political post of *protector de indios,* vigorously set about the destruction of Nahuatl cultural and religious materials, a goal not incongruous with his charge of protecting *indios* from Spanish excesses as well as their own blasphemous spiritual traditions.

d. The first school, opened in in Nezalhulpilli in 1523, instructed one thousand sons of noble Mexican and Texcocan families in the Latin alphabet, songs, arts and crafts, agriculture, and Christianity. Other schools were subsequently founded throughout Mexico, primarily driven by Franciscan, Dominican, and Jesuit religious orders.

FIGURE 2. *Mixity: Of Spain and India,* eighteenth century. Copperplate painting (48 x 36 cm). From Museo de America, Madrid. Photo by Album/Art Resource, NY.

connected Mexico City to San Juan Pueblo, in present-day New Mexico (United States). The Camino Real was the commercial lifeline of the colony, the route for silver extracted from the colony's mines to awaiting Spanish galleons in Veracruz and for linens, wines, tools, and other manufactured goods to the heartland of the territory. Transportation was initially carried out on the backs of indigenous slaves, but over time the Spanish shifted to oxen and mule trains, pulled by up to eighteen animals, which significantly increased the amount of cargo that could be carried.[67]

Precisely because roadways proved vital to extending the conquest into the northern regions of the colony, the diverse indigenous inhabitants, whom the Spanish referred to collectively as Chichimecas, launched frequent assaults on the wagon trains.[68] The natives welcomed themselves to the raided linens, wines, and food, but they found little use for the silver that was the inspiration for Spanish encroachment into their lands.[69] The Spanish efforts to quell the attacks came to be called the Chichimeca War, which lasted until 1600.[70] The eventual defeat of the Chichimecas did not ensure the security of mobility in New Spain, however. While banditry did not abound in colonial Mexico, as it would postindependence, population growth, unremitting unemployment, and the absence of political authority provided conditions ripe for its growth at the end of the seventeenth century.[71]

The existing legal body responsible for crime control in the colony—the Audiencia de Mexico and its Sala de Crimen (Criminal Court)—was powerless against the mobility and obscurity of bandits who preyed upon travelers. In response, the viceroy of New Spain took the radical step of founding a policing organization—the Tribuna de la Acordada—under the authority of Miguel Velázquez de Lorea. Velázquez assumed the post only on the condition that he be empowered to not only apprehend suspects but also adjudicate cases and execute sentences. By the 1780s, the Acordada was handling four-fifths of all criminal cases in the colony. Eventually, Velázquez's son, José, assumed command of the unit and founded a Guarda Mayor de Caminos (Elite Guard of Roadways) to root out roving bandits targeting travelers. The Guarda established guardhouses along major transit routes in the colony and escorted travelers as they passed.[72] Through these autocratic measures, the Acordada was slowly able to take control of Mexico's roads and lessen the threat of banditry to the colony's commerce.

As these examples from Mexico's past demonstrate, royal authorities were intimately aware of the importance of "enrolling"[73] material agents in their colonial project. The *sistema de castas, pinturas de casta,* and the construction and policing of roadways evidence how Spanish authorities crafted a framework to capture and cultivate the material dimensions of identification, communication, and mobility in New Spain to ensure the permanence of Spanish rule and protect it from the threats of foreign bodies, alien cultures, and banditry. And by doing so, central elements of the colonial state in Mexico—parochial schools and the Acordada, for instance—came into being.

2.3.3 *Building the State of Independent Mexico*

Following the expulsion of Spanish rule in 1821, Mexico faced a number of challenges to establishing itself as an independent country, including a reactionary European continent unsympathetic to revolutionary ideas following the defeat of Napoleon, the need to borrow heavily from other nations in order to bolster its military capacity against an expected Spanish invasion, a deeply entrenched Catholic Church unreceptive to the liberal ideas that inspired the independence movement, and an intense regionalism that prevented establishment of a strong national government.[74] As a result, independence brought a long spell of political instability to Mexico, punctuated by a Spanish attempt to reconquer its former dependency from 1821 to 1835, the loss of half of its territory to its bellicose neighbor to the north following the Mexican-American War from 1846 to 1848, a French-imposed monarch (Maximilian I) from 1864 to 1867, and dozens of national administrations in between.

This instability notwithstanding, independence and the embrace of Enlightenment ideals exercised an influence on personal identification in Mexico. Independence from Spain did not bring racial harmony to the newly autonomous land. Indigenous peoples continued to be seen as an impediment to progress, a view now sanctioned by the scientific veneer of social Darwinism.[75] But a notion of citizenship was developing in the country, which provided a broader basis for social integration than did Spanish rule. Both Benito Juárez, who served as president of Mexico for five terms, and Porfirio Díaz, who rose to prominence as Juárez's military chief, were of indigenous descent, and they and others in the ruling class saw *mestizaje* and the education of *indios* and the masses more generally as the path forward for the nation.[76]

Citizenship also implied voting rights. And independence from Spain initiated a period of experimentation with the democratic process that had profound impacts on personal identification and construction of the state. The first elections in postindependence Mexico operated under a "vague definition of suffrage that included no specific income or property restrictions."[77] Individual eligibility to vote was determined at local polling stations by election officials, a method of identification that could be exploited by those best able to mobilize the masses. In Mexico City, for instance, Masonic lodges vying for municipal power rounded up large numbers of people from poorer neighborhoods and supplied them with identical ballots with which to descend upon the city's eighteen polling places.[78] With no means for timely verification of voter identity, such a massing of people overwhelmed the system's ability to distinguish individual voters and resulted in electoral successes for the lodges.

Such maneuvers produced fears among the ruling elite that democratic governance would result in a loss of power to groups deemed either morally or racially inferior. In response, more restrictive definitions of citizenship and more precise methods for identifying citizens were developed. Voting rights were restricted to males eighteen years old and above (or below if married), employed (but not as domestic servants), and with no criminal records, public debts, or physical or moral incapacities.[79] This definition of citizenship not only excluded women but left criminal law as a key mechanism for excluding the poor and indigenous from voting.[80] Unsurprisingly, criminality intersected with categories of class, race, gender, and sexuality, which served to exclude from the democratic project those groups historically marginalized in Mexican society. Identifying who was left to vote was handled by local election commissioners, who conducted censuses of their precincts to generate registries of eligible voters. Voter rolls were to be completed eight days prior to elections, and they were also to be made public so that the citizenry could petition for inclusion. Ballots were then mailed to voters, which served as a voter's credential and identity document on election day.[81] In this way, democratic elections brought forth new methods of personal identification that are familiar to us today.

The political instability defining the first decades of Mexico's independence eventually gave rise to Porfirio Díaz, who held power in Mexico almost continuously from 1876 until the outbreak of the Mexican Revolution in 1910. Díaz's dictatorship, referred to as the Porfiriato, is generally read as a period in which strong militaristic rule forced political obedience upon society while erecting a framework for foreign investment and wealth extraction that exacerbated existing social inequalities and pushed the country to revolution. Critical to this period of political stability as well, however, was the development of modern systems of railway transportation and telegraph communication that allowed the central government to maintain ties to the state governors and local *jefes politicos* (political officials) it appointed to rule over Mexico's historically unruly regions (fig. 3).[82]

These transportation and communication technologies helped bring forth key institutions and a rule of law that are characteristic of modernity. As Teresa Van Hoy notes in a fascinating social history of the railroad, whereas Mexican governments before the Porfiriato handled railway development in an ad hoc fashion, which left its fate to investors and developers, Díaz increased the government's hand through such offices as the Inspector of Railroads in 1877 and the Communication and Public Works Secretariat in 1891.[83] In addition, the regime passed legislation, such as a land expropriation law in 1882 and the Federal Railroad Law of 1899, to provide land rights to railroad developers.

While these legal rules and political bodies were meant to increase the power of the federal government to seize land from people, they actually helped residents strengthen and defend their land claims against

FIGURE 3. Map of railway expansion during the Porfiriato. From William Henry Bishop, *Mexico, California, and Arizona* (New York: Harper and Brothers, 1900).

rail developers.[84] This unforeseeable outcome reveals much about how political power works in practice. To carry out the land expropriation law, the federal government sent officials to local communities to oversee railroad construction and report expropriation cases back to the federal ministry.[85] Once on site, however, officials encountered local authorities eager to use the railroad project to preserve their own power.[86] In the push and pull between local and federal-level politics, a new legal space was created, where residents were able to "maneuver for better terms than the government itself intended."[87] In the end, railroad development "inserted intermediaries into local power structures" and, in so doing, "linked local residents to federal authorities in ways that often channeled demands up more effectively than it imposed central authority down."[88] As a result, a "rule of law" began to take root over the "personalism" that had defined politics in Mexico for centuries.[89]

2.3.4 Realizing the Revolutionary State

The political stability of the Porfiriato did not last, as the economic inequality it engendered and the political abuses it relied on brought the country to rebellion. The Mexican Revolution represented a new era of civil strife and political chaos. And the near decade of intense fighting carried an enormous human toll and eviscerated the country's economy. Following the conflict, national leaders turned their attention to ensuring the peace by demobilizing armed groups, recuperating the country's agricultural production, and implanting revolutionary ideals across society. The general strategy for achieving these goals was crafted by Plutarco Elías Calles, president of Mexico from 1924 to 1928, who founded the National Revolutionary Party (PNR) in 1929, which would later become the Institutional Revolutionary Party (PRI) that held power until 2000. The corporatist structure of the PNR stabilized political and economic life by ensuring that political offices were determined by institutional forces rather than powerful individuals, corralling militant workers into state-sponsored unions and propagating revolutionary ideals through public schools and teachers who counteracted the power of the Catholic Church. The corporatist state model reached its apex under the administration of Lázaro Cárdenas (1934–40), who redistributed forty-five million acres of land from large landholders into communal holdings known as *ejidos* and nationalized the country's oil reserves in 1939.[90] In this way, somewhat ironically, "the formula of political centralism and dependent capitalist development" begun under Porfirio Díaz was perfected by the revolutionary leaders who had fought to erase his influence on national development.[91]

Practically speaking, the state's plans to redistribute land into *ejidos* and reactivate the agricultural sector posed special challenges. Based in Mexico City, federal officials remained largely ignorant, if fascinated, by rural life in the Mexican hinterlands. Thus, to carry out land redistributions and better plan the rural economy, the federal government sent out teams of data collectors and surveyors in its 1930 Agrarian Census to collect statistics and generate maps that would make this reality legible to the state.[92]

This was not Mexicans' first exposure to statistics.[e] But the devastation of the revolution and emergence of statistics as a basic language of

e. The Mexica, for instance, tracked tribute paid by subjugated rivals (Ervin, "1930 Agrarian Census in Mexico," 539). The Spanish Crown, too, when looking to check the growing power of *encomenderos* (land-grant holders), instituted a tax that required "a counting of family heads," which functioned like a census (García Martinez, "Los años de la conquista," 190). The implementation and improvement of accounting practices were a core concern of the Bourbons, who came to power in Spain in 1713 and sought to reverse centuries of apathetic management and tax collection in New Spain (MacLachlan, *Criminal Justice in Eighteenth Century Mexico*).

politics provided extra impetus to authorities' data-collection efforts in the 1920s and 1930s.[93] Communicating in the new language of statistics was not simply a matter of creating bureaucracies or sending young, idealist agronomists into the *campo*. Resistance met data collection at various points. Nonuniformity of measures was one. What constituted a "load," or *carga*, of wheat, for example, changed from locale to locale. And language made it worse. While statistics collectors found the "aztec-isms, or voices of Nahuatl origin, that live in the mouth of the Mexican farmer" to be problematic, Michoacán's "tarasc-isms," Yucatán's "mayanisms," and Oaxaca's "zapotec-isms" and "mixtec-isms" compounded the confusion of measurements. One report claimed that the "lack of unified terminology" was "particularly prejudicial" to state efforts in the agricultural sector.[94] What is more, the names of locales changed too, making it hard for officials to even locate places. Most vitally however, *campesinos* and large landholders resisted giving information to census takers, fearing the information would be used for the purposes of taxation or land appropriation.[95]

In response, the government and census workers worked to tie the new language of agrarian statistics to popular interests. Taking part in the census was "patriotic," "privacy" would be respected, "productivity" would increase and thereby benefit all, and "penalties" would be levied against those not participating. In addition, contests were held for best slogan to encourage Mexicans to participate in the census, balloons were distributed to schoolchildren with the word "Census" emblazoned on them, and teachers and state workers were enlisted to encourage those within their spheres of influence to take part. Despite these efforts, data collectors were ultimately left to negotiate the modernization project, trading information from farmers for the provision of water to an area. Still, these efforts helped bring into existence a modern state infrastructure to account for and communicate the material and social reality of Mexico.

With regard to mobility, the postrevolutionary state in Mexico sought to consolidate its power through the construction of roadways. Roads have a long history in Mexico, having served as the conduit not only for the extraction of mineral resources during the colonial era but also for the conduct of regional commerce during pre-Columbian times.[96] Attention to road construction and maintenance diminished somewhat during the Porfiriato, as rail travel was prioritized.[97] But with the growing importance of the automobile following the revolution, the federal government saw the construction of roadways as a way to tap

FIGURE 4. Workers operating machines in the construction of a roadway, c. 1925. Photo from CONACULTA-INAH-MEX, reproduction authorized by the Instituto Nacional de Antropología e Historia.

the potential of this new form of mobility and better incorporate rural areas that remained outside the state's grasp (fig. 4).[98]

Initially, the postrevolutionary state focused on constructing new roads that would access areas untouched by railroads.[99] A national highway system connecting Mexico's major cities and key economic areas soon followed. While foreign companies financed and constructed these new roadways at first, Plutarco Elías Calles, reflecting the nationalist inclinations of the postrevolutionary government, phased out their participation and founded the National Roads Commission in 1925 to oversee road construction and implemented a gasoline sales tax to finance the effort.[100] Around the same time, Calles nationalized all roads connecting the capital to key economic regions, state capitals, and neighboring municipalities.[101] And to better attract the foreign currency of US tourists who had begun to frequent the country's beach resorts, the government prioritized construction of roadways that connected northern points of entry to tourist destinations.[102]

By the middle of the twentieth century, the state's road construction strategy had evolved. With a growing Mexican automobile industry, the federal government dedicated 20 percent of all tax collected on the sale of vehicles assembled in the country to the construction of rural highways.[103] The prodigious growth of Mexico's cities, meanwhile, brought traffic jams, a problem addressed through the construction of ring roads around metropolitan areas.[104]

If the corporatist state model crafted by Calles and Cárdenas proved effective in creating an efficient highway system that integrated diverse regions of the country and supported Mexico's industrialization from the 1940s to the 1970s, it began to exhaust itself by the 1980s. The highway system was aging. And the geography of international trade had shifted, as increased commerce with Asia and economic growth of the US West created the need for new highways in the western region of Mexico. Public funding, however, had become insufficient to finance and maintain new roads. In response, President Carlos Salinas de Gortari (1988–94) launched the National Highway Project in 1989 to erect a new network of toll highways and bridges, which would be funded and operated by private companies.[105] Thus, roadways not only reflect authorities' efforts to harness automobility in modern Mexico, but they also trace the evolution of the postrevolutionary Mexican state from its corporatist origins in national integration and import substitution industrialization to its neoliberal embrace of global integration and export-oriented industrialization.

Finally, if elections in Mexico during the rule of the PNR/PRI were a sham, determined by the *dedazo* (fingermark), or personal preference, of the outgoing president rather than popular will, the hegemonic party held the democratic ideals of the revolution in high enough regard to at least hold elections. And elections occasioned the evolution of modes of personal identification for the purposes of determining voter eligibility. Following the revolution, the practice of mailing voters ballots that served as voter credentials and identity cards was discontinued. Instead, political parties prepared voter documents, which they sent to polling stations ahead of election day. This practice eventually gave way to the establishment of permanent voter rolls, which were compiled and verified at state and local levels. Following persistent charges of voter fraud, and reflecting the centralizing tendency of the postrevolutionary state, the federal government passed the 1946 Federal Election Law, which shifted authority over voter rolls from local officials to the Federal Elections Commission.[106]

The Federal Election Law also defined requirements for voter cards. They were to be created in numerical succession, with no repetition, and in duplicate, with the original given to the voter and the copy remaining with the commission. Thus, voters came to be distinguished by a unique identifying number. In 1951, the Federal Elections Commission further required voter cards to feature fingerprints as biometric proof of identity.[107] In 1987, additional elements were added to the card, including a security code and the imprinted signature of the general director of the National Registry of Voters.

These reforms did little to boost citizens' trust in the electoral system. The problem came to a head in the 1988 presidential election, when IBM computers responsible for counting votes famously crashed as opposition candidate Cuauhtémoc Cárdenas was leading the ruling party PRI-nominee Carlos Salinas. Salinas was ultimately named the winner. In the subsequent public outcry, the federal government created the Federal Electoral Institute (IFE), run by civilians and responsible for administering future elections. The IFE launched a nationwide census to verify voter rolls in 1991 and began issuing a new voter card in 1992, which included a photograph. As an individualized, accurate, and non-partisan form of identity, the IFE voter card became the dominant form of identification in Mexico.[108]

* * *

As these examples from the long arc of Mexican history show, the interest of authorities in controlling communications, identification, and mobility in Mexico is not new. Rather, such control is a constant goal of authorities, the pursuit of which gives shape to the state. The destruction of indigenous culture and imposition of the Spanish language in Mexico, the creation of the Acordada to police the Camino Real and other colonial roadways, the execution of surveys and use of ballots as identity documents during early independence, the establishment of the Inspector of Railroads and the Communication and Public Works Secretariat to oversee the construction of railways during the Porfiriato, the founding of public schools during the Calles and Cárdenas administrations, the launching of the National Highway Project by Carlos Salinas in 1989, and so on all represent efforts by state authorities to channel the collective agency of society, which in turn defined the shape of the state.

This aspect of state formation has been touched on by STS scholars exploring the boundaries of science, technology, and politics.[109] As Sheila Jasanoff observes, "The dynamics of politics and power, like those of culture, seem impossible to tease apart from the broad currents of scientific and technological change. It is through systematic engagement with the natural world and the manufactured, physical environment that modern polities define and refine the meanings of citizenship and civic responsibility, the solidarities of nationhood and interest groups, the boundaries of the public and the private, the possibilities of freedom, and the necessity for control"[110]

But this historical perspective offers more than a view of state formation through authorities' efforts to control collective agency. It also reveals a dialectical dimension of state formation, whereby arrangements of the state that once proved satisfactory in administering society lose their efficacy as society evolves through the introduction of cultural and technological elements that alter how communication, identification, and mobility are accomplished.[f] The extraction of mineral wealth in New Spain through the construction of roadways and establishment of shipping lanes lent the territory a level of activity that the Crown ultimately could not handle. Nearly all of the seaports by the end of the sixteenth century were riddled with pirates and buccaneers, and the mercantile system was flouted by Spanish subjects on both sides of the Atlantic.[111] Meanwhile, English, North American, and European traders pushed to open up new trade routes.[112] In the interior of the country, the continued construction of roadways rendered the guardhouses of the Guarda Mayor de Caminos of the Acordada ineffective. There was simply too much terrain for Spanish authorities to cover and too few resources with which to cover it. The advent of the railroad later provided the Porfiriato a novel manner for reordering the mobility of Mexican society and, in the process, developing the country's economic potential and the state's coercive capacity. But this mechanism ultimately proved incapable of quelling the societal unrest expressed in the Mexican Revolution and obsolete for constructing a postrevolutionary state in a world inhabited by the automobile. The national project of highway construction launched by the postrevolutionary governments of Elías Calles and Cárdenas proved effective in extending the reach of the central state to rural areas untouched by railroads and expanding trade with the United States. But the highway system began to lose its grip on mobility in Mexico once the corporatist state model experienced

f. I borrow this idea from STS scholars, who describe how scientific and technological production is forever introducing "actants" (Latour, *Pandora's Hope*) into society, nonhuman things with which humans can associate and amplify their capacity for action in the world. As a consequence, society is continually evolving in a state of "open-ended becoming" (Pickering, "New Ontologies," 4), where patterns of social interaction at one time do little to determine or predict patterns of social interaction or ways of being in the future.

financial crises, the industrial geography of the world shifted, and tourists began visiting the country via airplanes rather than automobiles. In sum, while authorities' efforts to capture the social processes of mobility, communication, and identification help give form to the state, the "vibrancy"[113] of society is forever challenging the modes for ordering it.

2.4 SURVEILLANCE TECHNOLOGIES AND REFORMING THE MEXICAN STATE

These observations on the dialectical nature of state formation help set the scene for more fully understanding the significance of the RENAUT, CEDI, and REPUVE today. If attempts to manage communication, identification, and mobility are part and parcel of the state, then the launching of the RENAUT, CEDI, and REPUVE speaks to the failure of prevailing state infrastructure to effectively administer these aspects of collective agency. There has been considerable debate in recent years about the crisis facing the Mexican state. Noting its inability to control crime, observers in the international press and political arena have been quick to conclude that Mexico represents a "failed state." On September 9, 2010, US secretary of state Hillary Clinton courted controversy when she opined that the insecurity in Mexico was "looking more and more like Colombia looked 20 years ago, where the narco-traffickers controlled certain parts of the country," an opinion that US president Barack Obama had to quickly distance himself from.[114] Scholars, meanwhile, noting the historical symbiotic relationship between organized crime and political authorities, are more likely to conclude that the federal government in Mexico remains very much in charge.[115] The truth, as always, lies somewhere in the middle. The increasing levels of crime in Mexico evidence neither a failed state nor a state in control, but rather a state that is progressively losing its grip on society.

Society is continually evolving in conjunction with cultural and technological elements that modify the way communication, identification, and mobility are accomplished. In contemporary Mexico, the increasing number of mobile phones, computers, digital networks, automobiles, and so forth has changed the way people communicate, self-identify, and move and has diminished the state's capacity for control. Law enforcement is unable to track the telephone communications of kidnappers, corpses are buried in mass graves without being identified, and criminals are able to circulate freely through society given the lack of traffic regulation. In this way, these material agents have altered and amplified collective agency in Mexican society, causing it to overwhelm the state's capacity to harness it.

However, the difficulties facing the Mexican state today reflect more than a society vitalized by infusions of technology and materiality. They also reflect wider processes of change since the 1980s. Some of this change involves external forces. For one, the global embrace of neoliberal political economy resulted in Mexico's abandoning the corporatist postrevolutionary state model that Presidents Calles and Cárdenas had forged in the 1930s. Economic liberalization was initiated by President Miguel de la Madrid in the 1980s and deepened with the North American Free Trade Agreement (NAFTA) in 1994. While economic liberalization has made some of Mexico's national industries more competitive, fostered trade with the United States and other countries, and led to a steady expansion of GDP,[116] the poverty rate (as measured by income required for basic living expenses) has remained stuck at 50 percent of the population[117] and the Mexican state has less ability to manage its effects.

Along with neoliberal political economy, the 1980s and 1990s saw Mexico and nearly all of Latin America undergo an era of democratization where competitive national elections were held for the first time in decades. In Mexico, the democratization process started at the local and regional levels, but it eventually resulted in the PRI losing the 2000 presidential contest. While this process has brought about free, competitive elections and increased civilian control over the political process, the resultant decentralization of power and weakening of clientelist relationships have diminished the power of the state to negotiate conflicts between competing social sectors.[118] And this has a clear relation to organized crime. Throughout modern Mexican history, the government turned a blind eye to, if demanding its wallets filled by, drug traffickers, so long as they kept to their own trafficking routes, or *plazas*, and did not engage in excessive violence.[119] With democratic elections and the PRI's loss of power, organized crime syndicates have turned the tables and are now able to dictate terms to politicians in exchange for campaign financing and other incentives.[120]

Also, globalization has transformed the economic and cultural landscape of Mexican society. The economic dimensions of globalization, which tie to neoliberal political economy, have resulted in increased foreign investment in sectors such as automobile and electronics production and have made information technologies such as personal computers and mobile phones more available. Connected to this, the opening of borders to facilitate international trade has created opportunities for international drug traffickers to take advantage of Mexico's immense border with the United States. The cultural dimensions of globalization have also had an impact on state control. At one time, the Mexican government used its monopoly over pulp to influence newspaper coverage in the country.[121] In a digital, networked world, however, its ability to control popular opinion has shrunk. And the infusion of global culture through foreign movies, television shows, and the Internet has transformed cultural values in the world's second most populous Catholic country.

Together, these external pressures have transformed state power in Mexico. In a sense, the balance has shifted. Neoliberalism, democratization, and globalization have shrunk the size of the state and diminished its ability to provide for people's economic welfare, to shape popular opinion, and to manage crime, while strengthening the hand of other actors, such as multinational corporations, domestic media companies, and drug cartels.

In addition to these external pressures, the power of the Mexican state has been diminished by internal pressures or contradictions. One example of this is the redundancy of state infrastructure, which the Citizen Identity Card is intended to address. As the modern state developed, distinct bureaucratic institutions arose around the different functions thought to belong to government. The Federal Electoral Institute emerged to manage democratic elections. So too did the Tax Administration Service, to oversee the collection of taxes; and the National Defense Secretariat, to manage the national defense of the country. Each of these institutions, as well as others, possesses its own identity cards, which has resulted in a proliferation of identities in Mexican society. If a relatively minor inconvenience in people's everyday lives, the redundancy of identification runs counter to the state's interest in homogenization and centralization.

Another internal challenge is corruption. Corruption has a long history in Mexico. However, its valence is different than in the United States or the Global North. During Spanish Hapsburg rule, the Crown sold public offices to fund its military campaigns and state affairs in Europe. Thus, the path to political power for Mexicans in the colonial state was through the purchase of offices. In present-day Mexico, police are crippled by low salaries and sometimes required to purchase their own guns and gear, a situation that leaves them susceptible to corruption.[122] Officers are expected to engage in rent-seeking activities, with the monies they procure sent up the chain of command to fulfill quotas established by their superiors.[123] Despite its functional aspects, corruption ultimately diminishes the power of the state because proceeds from a traffic ticket negotiated at the site of offense go into the pockets of the police rather than the coffers of the state, and paperwork that would document the event is never generated. This allows suspect vehicles to escape the grasp of the formal legal order.

Finally, and building on the last point, much of the state's data on the people and things they look to govern are erroneous, which further decreases its administrative hold on society. In my interviews with REPUVE administrators, data errors were a common theme. As one federal official explained to me concerning vehicle registrations, "If I crossed the registration database with that of stolen vehicles using a license plate number, it would give me up to fifty registrations. Someone didn't put the plate number down correctly. So, it wasn't credible that from one license plate number I would have fifty stolen vehicles." This harkens back to the inaccurate data provided by *campesinos* and the nonuniform agricultural measures used in Mexico in the early twentieth century that rendered the 1930 Agricultural Survey largely useless in imagining "the country's agricultural realities."[124] Without a common set of reliable information, state authorities find it difficult to carry out the task of governing.

Thus, the current crisis in governance in Mexico reflects a period of transformation in which the society that authorities seek to order has become increasingly dynamic while the state has atrophied. Globalization has helped transform Mexican society by introducing material artifacts (automobiles, computers, mobile phones, digital networks) that alter the way people communicate, self-identify, and move, while neoliberalism and democratization have forced the state to surrender the corporatist model that once controlled society. The Mexican state is not today failed. Nor is it in control. Rather, it finds itself in a situation where it must reinvent its capacity for governing in order to tame the tiger of Mexican society.

This, in the end, brings us back to surveillance technologies and their significance in Mexico today. The RENAUT, CEDI, and REPUVE are attempts by the Mexican state to adjust the mechanisms of power to fit the conditions it now faces. Confronted by a state infrastructure riddled with redundancy, corruption, and error, governmental authorities turn to incorruptible, precision technologies to reduce their reliance on over-abundant, unprincipled, erring state employees.[125]

The Citizen Identity Card responds to the redundancy of identity in Mexico by shepherding the different identities of Mexicans developed over time into a single document. The Spanish patterns of naming persons established during the conquest are inscribed onto the card. The enumerated assignment of identity to Mexicans through the Unique Population Registry Code is added to the card via a bar code. The biometric means for identifying people inherited from the voter card—the fingerprint and photograph—are supplemented by iris scans. As a technological device, the uniqueness of the CEDI lies not so much in its use of biometrics but in its amenability to diverse methods of identification.

With the REPUVE, meanwhile, human cops who rent-seek by overlooking traffic infractions are "displaced"[126] by teams of RFID stickers and scanners that have no motivation for financial gain. Ignacio Meza, a state official responsible for implementing the REPUVE in Zacatecas, alluded to this same point when I interviewed him. He remembered a time, "a few administrations ago," when the state operated a program that sought to put the vehicle roll in order.

> The program was called Secure Transit. It was primarily a public safety program, but it also contained provisions allowing Finances to verify payments due. So, two Transit officials would go out, one person from Finances, and another person from the State Comptroller, who would serve as an observer to prevent acts of ill gains, abuse of authority, and so forth. This program allowed Transit authorities to monitor pollution emissions, tinted windows, seat belts, and so on. . . . The program helped elevate levels of compliance. But if every vehicle had a chip, we could start a new program of this type. With the scanner and chips, we could collect information on payments, and this would help the operation of such a program a lot.

It is worth noting that the older program required four people to operate, one to explicitly monitor the other officials. With the chips and scanners, another program could be launched, but without the need for so many human actors. Wondering how corruption would be affected by the program, I asked Meza what would happen with bribes. "Well," he replied, "we would be putting an end to that." Thus, by "delegating" to technological devices police work traditionally belonging to human law-enforcement officers, the state imagines itself more able to enforce vehicular regulations.

These considerations help mark out the true novelty of surveillance technologies in Mexico. The Mexican state's launching of the mobile phone registry, national identity card, and electronic automobile registry reveals an effort to reinvent itself in order to gain a handle on a rapidly changing society that is escaping its grasp. From this point of departure, what we need to consider in greater detail is how these programs might remake the state. Or, in other words, how are the programs designed to tame the tiger?

CHAPTER 3

Prohesion

We have wanted to show what is the only end awaiting the
criminal.

—*El automóvil gris*

The Public Registry of Vehicles (REPUVE) has the objective
of guaranteeing public security through the provision of legal
certainty to those acts carried out with vehicles circulating
within national territory.

—REPUVE white paper

3.1 EL AUTOMÓVIL GRIS

One of the most successful films of the Mexican silent era premiered
on December 11, 1919, at twenty movie houses in Mexico City. *El
automóvil gris* (The gray automobile) depicted the story of the Gray
Automobile Gang, a group of prison escapees who terrorized Mexico
City's elite at the height of the Mexican Revolution in 1915 by donning
uniforms of the Constitutionalist Army and serving false search warrants to rob people's homes. Originally twelve episodes, the shortened
version of the film that survives today is split evenly between drama-
tizing the gang's various robberies, kidnappings, and murders and its
eventual apprehension by the inspector Juan Manuel Cabrera, who por-
trays himself in the film. Imagined as a work of cinema verité, broad-
casting events of contemporary concern to Mexican society at a time of
immense social upheaval, *El automóvil gris* was a sensation, highlighted
by its notorious final scene, actual footage of the execution of Gray
Automobile Gang members by firing squad.[1]

Over the years, *El automóvil gris* has captured the imagination of
moviegoers, film buffs, students of Mexican history, and scholars from

different fields by preserving a unique moment in Mexican history. But while the film was promoted on the basis of its authenticity, it is better described as a "spectacle of representational authenticity."[2] In other words, *El automóvil gris,* the movie, takes considerable artistic license with *El automóvil gris,* the story. The film was financed by Pablo González Garza, a general in Venustiano Carranza's Constitutionalist Army, sent to pacify Mexico City following its occupation by Pancho Villa's and Emiliano Zapata's armies. González, who would later orchestrate the assassination of Zapata, is portrayed in the film passionately exhorting the police forces to pursue the gang. During the Gray Automobile Gang's reign of mischief, however, it was widely believed that González was the operation's mastermind. Whatever that claim's veracity, the film glosses over military and police complicity in the crimes of the gang, which served search warrants signed by actual military authorities.[3] The film thus functions as a political tool. Produced by a military general of dubious repute, the film attempts to rehabilitate his image and legitimize the nascent Mexican state by portraying the military and police as honest, hardworking protectors of the social order.[4]

For the purposes of the present work, *El automóvil gris* proves interesting for what it reveals about the Mexican elite's view of the automobile when the film came out. The motorcar, still a novel technology, was seen as a threat to the social order. Combined with telegraphic communications and a lack of clear identifying documents, it empowered wrongdoers to prey on the public and stay one step ahead of the law. Only through the heroic efforts of honest authorities such as General González Garza, Inspector Cabrera, and police officers skilled in the use of modern transportation and communication technologies could the criminality cultivated by the motorcar be controlled. "A useless endeavor," the film confidently professes at its conclusion, "the fate [execution by firing squad] awaiting all guilty men is a moral lesson that work is the only noble path in life."

But the insecurity of the automobile in Mexico was not so easily tamed. Today, the car remains a threat to security. Its value as a manufactured object makes it a frequent target of thieves.[5] Its mobility is critical to drug trafficking to the United States as well as arms to Mexico, attacks by drug cartels against other cartels as well as the police and military, and abductions. Its propensity for immobility and congestion is exploited by cartels to create roadblocks to impede police and military forces during shootouts.[6] Its enclosure of space provides the cover necessary for illicit trafficking and hiding corpses for authorities

FIGURE 5. Ciudad Administrativa in Zacatecas, 2013. Photo by Benjaz1213.

to uncover. And its combustibility has been used on at least one occasion to produce car bombs to punish state authorities seen to favor one cartel over another.[7]

The federal government's response to this insecurity—the Public Registry of Vehicles (REPUVE)—is currently on display in various states across the country. One of Mexico's first REPUVE registration sites was in the north-central state of Zacatecas. In the eponymous capital, named a UNESCO World Heritage Site in 1993 for its rich colonial architecture and urban layout, the REPUVE module can be found in La Ciudad de Gobierno, or Government City (fig. 5), a brand-new complex of shiny governmental office buildings on the city's outskirts, a location intended to simultaneously facilitate citizens' completion of their civic obligations and bolster the city's tourism by directing bureaucratic traffic out of the historic downtown. Within the impressive complex, the REPUVE module—housed in two semitrailers outside the Secretariat of Finances building—casts a humble shadow. Armed with computers, satellite antennae, and other advanced information technologies, these trailers comprise a starting point in the state's fight against automotive insecurity (fig. 6).

FIGURE 6. REPUVE registration site in Zacatecas, 2011. Photo by Keith Guzik, © 2015.

Citizens concerned about the safety of their vehicles come to the module for RFID tags, or *chips* as they are called colloquially, which enable the location of their vehicles in case of theft (fig. 7). To receive a *chip,* a driver must present five forms of documentation: a bill of sale from the purchase of the vehicle, a vehicle registration, a driver's license, a proof of residency, and an official form of identification—which data specialists in the trailers use to verify the driver's and vehicle's history.

Outside the trailers, REPUVE technicians inspect the vehicle, photographing its front, rear, and sides and locating three instances of its vehicle identification number (VIN), one of which is recorded via an impression made using transparent tape and chalk. After the driver's documents have been reviewed, and the identity of the vehicle verified, the data operators enter the vehicle's details in the computer and print out the RFID tag, which contains the vehicle's VIN, the tag's identification number, and the corresponding REPUVE file number. One of the technicians then applies *el chip* to the inside windshield above the rearview mirror, and the technicians then test it using a handheld RFID reader (figs. 8–15). The registration process complete, the tag can now be registered whenever the vehicle passes REPUVE readers installed along roadsides and other transit points. In this way, the fight against automotive insecurity in Mexico is transferred from the heroic police officers of *El automóvil gris* to the mundane surveillant technologies of the Public Registry of Vehicles.

This chapter focuses on the tension between the disruptive collective agency engendered by the automobile and the efforts of Mexican

FIGURE 7. REPUVE RFID tag. Photo by Keith Guzik, © 2015.

authorities to control it. It describes how automobility[8] has served at different times in Mexican history as an engine of insecurity, threatening the physical safety of motorists and the ecological welfare of the natural environment. In response to the insecurity produced by the motorcar, authorities have turned to the law to keep their hands firmly on the wheel. At the dawn of the automobile in early twentieth-century Mexico, the administration of automobiles focused on ensuring personal safety (and collecting taxes) by disciplining motorists to be responsible drivers. To do this, the state imposed three basic requirements for the operation of a motor vehicle: *registration* of both cars and drivers into state registries, *inspection* of the functionality of automobiles and the competence of drivers, and *regulation* of motorists' compliance with traffic rules.

Beginning in the 1980s, a new legal regime emerged that sought to reduce the pollution "risk" of automobiles. In this model, *registration* still recorded the correspondence between cars and drivers, but emissions testing now recorded a mostly invisible discharge from cars made visible during *inspections* through diagnostic machinery and translated onto vehicle surfaces through inspection stickers. *Regulation* under this risk model of automotive governance did not involve overseeing vehicles

FIGURE 8. Initial review of documents during REPUVE registration. Photo by Keith Guzik, © 2015.

FIGURE 9. Photographing vehicle during REPUVE registration. Photo by Keith Guzik, © 2015.

FIGURE 10. Locating VIN number during REPUVE registration. Photo by Keith Guzik, © 2015.

FIGURE 11. Recording VIN number during REPUVE registration. Photo by Keith Guzik, © 2015.

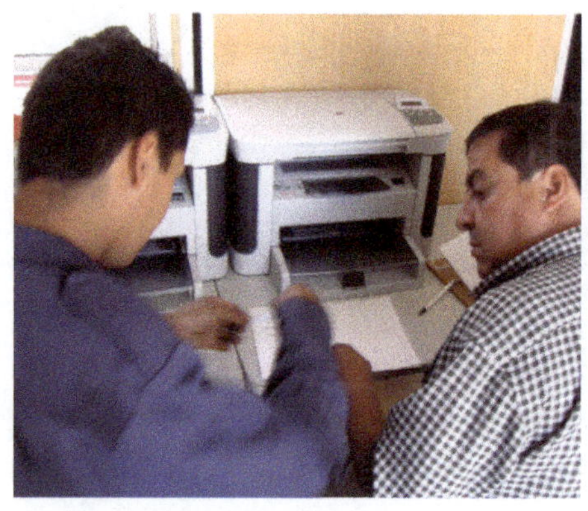

FIGURE 12. Transferring VIN number during REPUVE registration. Photo by Keith Guzik, © 2015.

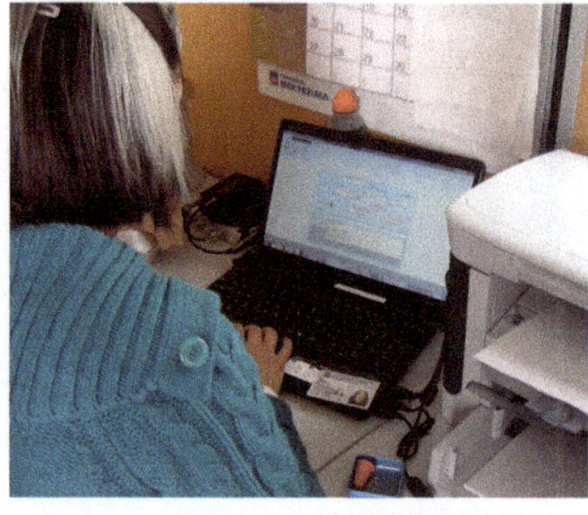

FIGURE 13. Inputting driver and vehicle data during REPUVE registration. Photo by Keith Guzik, © 2015.

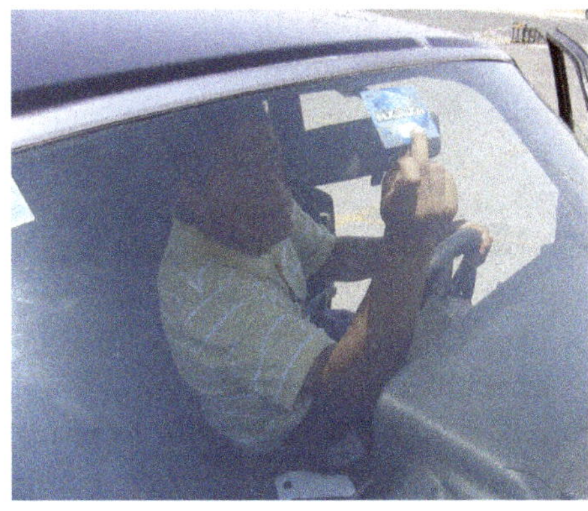

FIGURE 14. Adhering RFID tag during REPUVE registration. Photo by Keith Guzik, © 2015.

FIGURE 15. Verifying RFID tag during REPUVE registration. Photo by Keith Guzik, © 2015.

to ensure drivers' compliance with traffic rules, but rather supervising them to ensure drivers' abnegation from an activity—automobility—posing a risk to the natural environment.

In the REPUVE, meanwhile, the concern of the state is not on personal safety or environmental pollution, but "legal certainty" *(certeza jurídica),* the assurance that the automobile possesses proper legal status. To achieve certainty, a *registration* is made not only of the correspondence between cars and drivers, but of all records concerning a vehicle generated by public and private entities administering automobility, a massive integration of data made possible by the REPUVE database. *Inspections,* for their part, are not directed to the functionality or emissions of vehicles but to their identity as established by vehicle identification numbers carried on vehicle bodies. Finally, *regulation* is conducted not under the watchful gaze of human traffic officers but via the transmission of digital signals received by RFID tags adhered to vehicles at inspection sites.

In the REPUVE, then, one discerns a distinct mode of governmentality. If personal safety was pursued by creating a registry of car-driver couplings and overseeing the conduct of machines and drivers (an operation concerned with the conduct of automobiles), and environmental pollution was combatted by diagnosing vehicular emissions and producing visualizations of them for others to see (an operation focused on the essence of automobility rather than its conduct), then "legal certainty" is sought by integrating data about vehicles and adhering that data to the vehicles themselves (an operation that moves the power of the state from the conduct and essence of vehicles to the body of automobility). This double operation at the heart of the REPUVE—of making the multiple records of vehicles more *cohesive* and having those data become *adhesive* to the materiality of automobility—is what I call *prohesion,* an emergent logic of power distinct from traditional surveillance and discernible in other security measures in contemporary Mexico, where authorities attempt to control collective agency by making its material more amenable to governance. By reducing the state's dependence on unreliable, corruptible human subjects and increasing its use of more objective surveillance technologies, prohesion offers the prospect of increasing the state's control over the disruptive collective agencies at the heart of insecurity in Mexico.

3.2 AUTOMOBILITY IN MEXICO

The question of when the automobile appeared in Mexico is a matter of speculation. A motorcar was first officially recorded in the country in 1898, a handmade French machine named the Dellanu Villeville, which the Texas millionaire Manuel Cuesta purchased and transported to Guadalajara.[9] There are accounts of another motor vehicle rambling its way through the dusty streets of Mexico City in 1895, commanded by one Fernando de Teresa, but no official records exist.[10] Regardless of their origin, the first automobiles in Mexico found themselves competing for space on roadways with both more traditional modes of transportation (mule trains, horse-drawn carriages) and more modern ones (steam and electric railcars).[11] Those experimenting with the new form of locomotion had substantial financial resources. An imported automobile was a primary status symbol for upper-class families that marked their separation from the rest of society.[12] During the early days of the automobile in Mexico, only 215 units were sold per year.[13]

The number of vehicles in Mexico soon began to rise, however, from roughly 8,000 units in 1920 to 105,470 in 1940.[14] By the mid-1950s, the country counted nearly 500,000 vehicles, second only to Brazil among its Latin American neighbors.[15] The car's growing popularity involved more than the curiosity of the moneyed classes. One of the chief technological achievements of Porfirio Diaz's rule in Mexico was the electrification of Mexico City's rail system, which ably served the urban population's transportation needs. Strikes began disabling the system in 1916 and 1917, however, fueling the need for a commuting alternative. As a personalized mode of transportation not immediately subject to labor strife, the automobile offered a more reliable mode of transportation.[16] Outside the nation's capital, meanwhile, the national economy had long been challenged by geography. Without large rivers that could accommodate the transportation requirements of the country's major industries, Mexico proved a prime location for the growth of trucking.[17] Finally, a shared border with the world's leading manufacturer of automobiles provided the country with an ample source of vehicles.[18]

In his seminal work on the role of the automobile in modern society, John Urry uses the term "automobility" to describe the system of interrelated institutions, industries, historical processes, cultural practices, and emotions that have arisen around the automobile. In modern and postindustrial society, he writes, the car stands as

1. the quintessential *manufactured object* produced by the leading industrial sectors and the iconic firms within 20th-century capitalism . . . ;

2. the major item of *individual consumption* after housing which provides status to its owner/user through its sign-values . . . ;

3. an extraordinarily powerful *complex* constituted through technical and social interlinkages with other industries [car parts, road building, advertising, oil production] . . . ;

4. the predominant global form of "quasi-private" *mobility* that subordinates other mobilities of walking, cycling, travelling by rail and so on . . . ;

5. the dominant *culture* that sustains major discourses of what constitutes the good life . . . ; [and]

6. the single most important cause of *environmental resource-use.*[19]

The impact of automobility on Mexican society is unmistakable. The automobile's promise as a manufactured object has driven the country's economic policy since the first half of the twentieth century. While the first cars arrived in Mexico across the northern border with the United States,[20] the postrevolutionary government of Plutarco Elías Calles recognized the automobile's potential to advance industrialization and made development of a car industry a centerpiece of its import-substitution economy.[21] Mexico's large population already promised manufacturers a healthy market, and the country's substantial oil reserves, privately held until Lázaro Cárdenas's nationalization in 1938, provided the raw materials necessary to fuel cars and construct highways.[22] But to increase the country's appeal to foreign manufacturers, the Calles government offered favorable trade tariffs for companies establishing operations in Mexico and assurances that the inexpensive labor force grouped into party-affiliated unions would not disrupt production. Ford was the first producer to build a plant in Mexico, in 1925, receiving a 50 percent reduction in import tariffs.[23] Other manufacturers followed. By 1946, ten companies were assembling cars in Mexico, in plants located largely in and around Mexico City.[24]

While these operations met the government's immediate interest in establishing an automotive industry, the companies operated solely as assemblers, using knockdown kits produced in the United States to build cars for sale in Mexico.[25] This arrangement not only limited the extent of industrialization within Mexico but also led to a trade imbalance with the United States that deprived Mexico of needed foreign currency.[26] In response, the Mexican state began exerting increased control over the industry, primarily through import restrictions, local-content requirements, and price controls. In 1950, a government executive order imposed maximum prices on wholesale and retail items.[27] At the end of the decade, import quotas were reduced and the use of locally produced parts was decreed for the first time.[28]

Given persistent problems with balance of trade, the government moved in the 1970s to increase exports from the automotive sector. Companies located in Mexico and exporting to the United States were offered substantial subsidies to help offset US import tariffs.[29] In 1977, an executive decree required each producer in the country to balance its automotive trade and placed strict limits on foreign-owned investments not dedicated to export production.[30] During this time as well, the first *maquiladoras,* or assembly plants, were established in northern Mexico to produce auto parts for export.[31] In time, the number of *maquiladoras* in the automotive sector increased, owing to the reduced presence of labor unions, lower transportation costs to the United States, and ease of access to customs offices.[32] As a result, the number of automobiles exported from Mexico grew from 2,000 in 1972 to 58,423 in 1985, with Volkswagen and Nissan serving as industry leaders.[33]

These efforts succeeded in establishing an active automotive industry in Mexico, especially in parts production. But the statist import-substitution model of economic development began giving way in the 1980s to larger forces in international political economy. A rising yen, in the case of Japanese producers, and the need to reduce costs to remain competitive with Japanese producers, in the case of US and European companies,[34] led automotive multinationals to move more labor-intensive aspects of production to industrializing countries like Mexico. Attempting to adjust policy to fit the shifting world economy, the Mexican government as part of North American Free Trade Agreement (NAFTA) negotiations at the end of the 1980s, eliminated local-content requirements and rules governing foreign ownership of production plants.[35] These policy changes helped reestablish Mexico as an attractive base of operations for auto producers. Today, Mexico ranks eighth in the world in number of automobiles produced, around 2.9 million units per year,[36] and second in exports to the United States.[37] Thus, over the course of the twentieth century, Mexico transformed its relationship to the automobile as a manufactured object. Solely an importer at the start of the century, the country at its conclusion stood as one of the world's largest exporters, transforming its image as a backward, rural society to one at the cutting edge of the global economy.

Automobility has affected Mexico in ways beyond industrialization and consumption. At the most basic level, the motorcar has transformed time and space. Whereas urban life was once concentrated in city centers, the automobile enabled the growth of well-to-do suburbs on urban peripheries. Between 1920 and 1960, the urban area of Mexico City grew ninefold, and the very notion of the city changed. Formally comprised of sixteen *delegaciones,* or boroughs, *el distrito federal* has expanded to include over forty municipalities in the states of Mexico and Hidalgo, a conurbation referred to as the metropolitan zone, the urban area, the urban area of Mexico City, or the metropolitan zone of the Valley of Mexico.[38] The automobile not only expanded the boundaries of urban areas but also split them apart through its associated highways and central arteries. And as much as this spatial growth owes to the time ostensibly saved by the automobile in traversing long distances, this time is ultimately lost in the uncompensated hours and energy that people spend sitting in traffic.[39]

Similarly, the constant din of automobility in urban areas keeps people tense and disrupts sleep.[40] Not only is silence forgotten, but when it is encountered, in rural areas for instance, it provokes anxiety rather than calm.[41] And behind the wheel, drivers become obsessed with arriving at destinations in the least amount of time possible, a mind-set that transforms other vehicles, bicyclists, and pedestrians into obstacles to overcome.[42] In these ways, the automobile has transformed the national character of Mexico.[43]

The influence of automobility has extended to language as well. As Federico Fernández Christlieb notes in his excellent history of the automobile in Mexico, "After money and sex, no object has given birth to as much popular vocabulary and slang as the car."[44] Fernández offers *luz verde* (being given the green light) and *en curva* (being thrown a curve) as examples. But a more illuminating phrase from Mexican Spanish might be *tantas curvas y yo sin frenos* (so many curves and I don't have brakes). This crude expression combines the linguistic influences of both sex and the automobile, revealing national, patriarchal tendencies to objectify women (by comparing them to the physical landscape) and excuse men for improper conduct (by explaining away their lack of self-restraint).

The automobile, in this sense, is not merely a means of transportation, but a means of cultural representation central to Mexican national identity. Films—such as *Mecánica nacional,* a boorish 1971 comedy following the weekend excursion of a Mexico City repair shop owner's family to an automobile race, and *Y tu mamá también,* the first international hit of Alfonso Cuarón, a coming-of-age tale that also centers around a road trip from Mexico City—use the automobile to reflect on the state of Mexican society. The humble Volkswagen Beetle, meanwhile, produced in Puebla until 2003 and lovingly referred to as *el vocho,* has found itself the inspiration of leading contemporary artists from Mexico. Damián Ortega's *Cosmic Thing* (2002) installation features a disassembled Beetle suspended in air (fig. 16), Betsabeé Romero's works frequently center on Beetles (fig. 17), and Margarita Cabrera's *Yellow Bug* (2004) offers a life-sized, soft sculpture of the iconic car (fig. 18). These works speak to the central place of the automobile in representing everyday life in Mexico. *El vochol,* an actual Beetle fully decorated in the beads representative of Huichol art, an indigenous group from northern Mexico, takes this idea a step further, integrating the automobile into the cultural expression of a people known for their nomadic lifestyle (fig. 19).

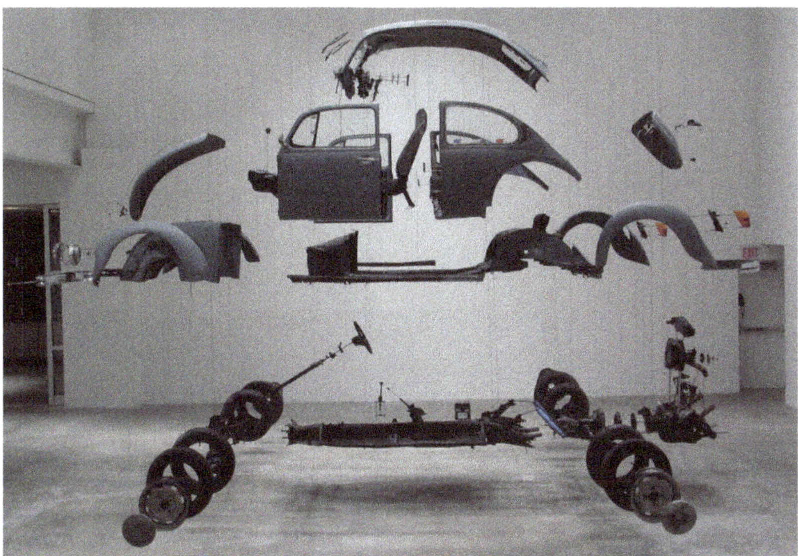

FIGURE 16. *Cosmic Thing*, 2002, by Damián Ortega. Image courtesy of the artist and Kurimanzutto, Mexico City.

FIGURE 17. *Arquitectura rústica para carros inseguros*, 2003, by Betsabeé Romero. Image courtesy of the artist.

FIGURE 18. *Yellow Bug*, 2004, by Margarita Cabrera. Image provided by Art © Margarita Cabrera/licensed by VAGA, New York.

FIGURE 19. *El vochol*, 2010, by eight anonymous Huichol artists. Courtesy of Asociación de Amigos del Museo de Arte Popular (AAMAP). Photo by Keith Guzik, © 2015.

Beyond the automobile's cultural dimensions, its increasing dominance in Mexico has come at the expense of other uses of public space and modes of transportation. It is, as John Urry writes, "the predominant global form of 'quasi-private' mobility that subordinates other mobilities of walking, cycling, travelling by rail and so on."[45] As noted earlier, trains powered first by steam and later by electricity and carriages pulled by beasts of burden capably served the transportation needs of Mexico City's population throughout the nineteenth century. But the postrevolutionary government's promotion of the automobile swept these modes of mobility aside. And when the problem of traffic congestion made urban railways desirable again in the late 1960s, they were primarily built underground as subways in accommodation of motorways.[46] More recently, the supremacy of the automobile has been questioned somewhat in Mexico City, as municipal authorities have embraced the expansion of regional trains, the establishment of bus-only lanes on major arteries, and the creation of a bike-share program to provide city residents with transportation alternatives. But for the time being, these examples can better be read as exceptions that prove the rule of automobility in everyday Mexico.

3.3 GOVERNING THE INSECURITIES OF AUTOMOBILITY

As comprehensive as the concept of automobility is in highlighting the multiple dimensions of modern life affected by the motorcar, it overlooks security and law. The oversight is somewhat curious, as research on the history of the automobile in the United States describes how its introduction led to a spate of traffic accidents,[47] endangered children playing in the streets,[48] increased the risks of rape for women,[49] gave birth to "roving criminals,"[50] and threatened public safety through intoxicated drivers.[51] In Mexico, likewise, the automobile has consistently served as a source of insecurity. The insecurity of automobility has a dialectical, evolutionary quality, whereby the dangers of car travel in one time period give rise to state interventions designed to address them, which eventually cede ground to new perils. This section traces this progression across three time periods over the last century.

3.3.1 Responsibilizing Motorists

Government documents and other literature from when the automobile first appeared in Mexico give insight into the insecurity of automobility.

A 1938 pamphlet titled *Seguridad!* (Security!), published by the Secretariat of Communications and Public Works, describes the different types of accidents involving automobiles on federal highways: 1,273 incidents were reported in 1937, among them 15 fires, 61 crashes against embankments, 125 people run over, 369 collisions with other vehicles, and 508 overturned vehicles, all of which claimed 778 lives.[52] These are not large numbers. Today, Mexico reports 16,700 road fatalities per year.[53] But when one considers that these numbers do not include unreported accidents or accidents occurring on state and municipal roadways, and that Mexico was home to far fewer vehicles at this time—105,470 in 1940[54]—the risk to personal safety captured by these figures is not insignificant. The fatality rate of 778 deaths per 105,470 vehicles in 1937 works out to 737.7 deaths per 100,000 vehicles. The rate today in Mexico is 54.1 deaths per 100,000 vehicles.[55] It is not difficult to understand, then, why the secretariat in publishing these numbers warned that, without proper attention, automobile travel would be "converted into an epidemic."[56]

Seguridad! also gives the causes of these accidents. In the 1,273 incidents, there were 39 cases of headlight problems, 58 of inexperience, 69 of carelessness on the part of pedestrians, 68 of drunkenness, 169 of traveling in the wrong lane, 108 of poor vehicle maintenance, 149 of excessive speed in turns, 176 of excessive speed in corners, and 441 of lack of precaution.[57]

What stands out is how a lack of precaution on the part of motorists, whether explicit or implicit in the categories listed, is seen by the government as the main culprit behind the insecurity of automobiles. Guilt accrues to the individual, which suggests that the threat of automobiles to personal safety could be curtailed by having drivers take better care when operating motor vehicles. This is not to say that the government at the time was blind to other possible causes of automotive insecurity. Manufacturers had a role to play by "offering drivers vehicles in perfect conditions . . . with steering responsive to the smallest impulse of the driver, breaks with unquestionable efficacy, powerful headlights, proven road stability, and, in a word, equipped with all the elements to fulfill its function with a minimum of risk."[58] Public authorities could also do more to "increase visibility, improve curvatures and gradients in roadways, repairing them constantly and seeing to their conservation, placing signals in dangerous spots to help drivers, and launching preventative measures as well as punitive ones when needed against transgressors."[59] Nevertheless, in the end, the secretariat concluded that the most effective way to fight the insecurity of automobility was to resolve the "problem of educating drivers and pedestrians."[60]

To educate motorists, governmental agencies and other organizations interested in promoting automobiles released publications on proper motorcar usage. Some of the instructions were elementary. Drivers, for instance, were instructed that "in crossroads, the traffic rules dictate that the car arriving to our right has the right-of-way, and the same goes for street crossings."[61] "When driving at night, always keep your headlights and rear light illuminated."[62] And "one can never trust the sides of the road in partial or total darkness."[63] Other instructions seem quaint today: "Never under any circumstances complete a turn using sidewalks, but rather use the center of the crossing."[64] Also, "A lack of precaution is lighting a cigarette in a gas station, since gasoline is not only given via pumps, but also with cans and jars."[65]

These publications, in addition to instructing motorists on the rules of the road and proper driving technique, also sought to prepare readers for the emotional challenges of driving. The *Manual del "chauffeur"* (Driver's manual), published in 1943, explained that "it often happens that drivers . . . throw themselves into a crazy race to catch and overtake a car that has committed an 'offense' against them. If the driver of the other car becomes aware of what is happening, a true speed contest tends to result, with a total lack of respect for others on the road as well as oneself. One should realize the absurdity of such behavior, meant only to satisfy a stupid vanity."[66] "It should never be forgotten," the manual admonished, "that we do not own the road; other users have as much right as we do."[67] Drivers were urged to keep vehicle maintenance in mind too. "Accidents occurring because of excessive speed do not always indicate a lack of skill on the part of drivers," noted the pamphlet *Seguridad!,* "since many have already demonstrated their ability to drive, [but] many times the unexpected happens, brake failure or a tire that blows or falls off."[68] As a result, "before beginning an automobile trip, the driver should ensure that the car is in the right conditions, and to do this, it is appropriate to examine it quickly and verify that all its parts function without problems."[69]

In addition to information campaigns, government authorities combatted the insecurity of automobility by establishing legal requirements for their operation. Although not the first government entity in Mexico to do so, the City Council of Ciudad Juárez had already published its *Reglamento de vehículos* (Vehicle rulebook) in 1922. The stated purpose of the book was that "all property of this type"—vehicles—"be registered, inspected, and regulated." "Vehicles" at this time was an inclusive term. The registry applied not only to automobiles but also horse-drawn carriages and other types of hotel, commercial, and transport cars.[70]

The rulebook's statement of purpose was an effective summary of the general requirements for operating a vehicle in the city. Vehicle owners first needed to go to the treasury to register their machines, done by making "a written request to the treasury explaining its [the vehicle's] purpose, presenting their property to the local inspector for inspection and classification, and promising to follow the rules established in this rulebook." Then, "following authorization by the inspector, the applicant should appear before the municipal treasury to request a tax book, in which payments in accordance with this rulebook will be noted, understanding that it is required to complete these payments within the first eight days of each month." For vehicles serving the public, such as taxis, the registration requirements were stricter. "Once the application is approved," the rulebook explained, "the applicant should request the corresponding registration number in order to mark his property thereafter, obtain at his own cost a copy of the respective fee and a copy of this rulebook. In the rulebook noted, there should appear as well a drawing of the applicant's bust, noting at the bottom his appearance, personal data, residence, names of persons who recommend him or would testify for him before the municipal president, and several pages left blank in the back of the book to note inspections and other particularities that the police find necessary."[71]

The second stated goal of the rulebook was vehicle inspection. Such inspections were presided over "by the city councilman and the vehicle inspector." Drivers were required to "present at the inspection all of their personal documentation, including the tax book." And "all public-service vehicles" were required to "pass a monthly inspection." Drivers also had to be inspected. To qualify to drive a vehicle, individuals were required to "be expert and experienced in the conduct of the machines, as determined by a jury consisting of the inspector, a witness provided by the applicant, and a third party named by the municipal president; be eighteen years old; and have good morals."[72] Finally, once inspected and registered, "the automobile should display both in the front and rear, clearly visible and no less than 5 inches in height, the order number corresponding to it in the registry."[73]

In addition to registration and inspection, the municipal government of Ciudad Juárez sought to ensure that automobiles followed "the rules established in this rulebook." Some traffic rules have been mentioned above—"Never under any circumstances complete a turn using sidewalks, but rather use the center of the crossing" and "When driving at night, always keep your headlights and rear light illuminated." The traffic code also established speed limits: 7.5 miles per hour (12 kilometers per hour, or kph) within the city, 4.4 miles per hour (7 kph) on side streets, 3.1 miles per hour (5 kph) in front of schools, 1.9 miles per house (3 kph) around public promenades, and 15.5 miles per hour (25 kph) "away from populated areas in the most adequate and least dangerous stretches."[74]

In establishing the rules of the road, the city council considered more than personal safety. Public order and morality were also at stake, especially in vehicles used for public transportation, where the class divisions of Mexican society would play out. The Ciudad Juárez rulebook dictated that "when driving drunks or prostitutes, drivers should conceal their vehicles with their covered roofs, driving them directly to their destination." "Calling out to passengers by whistles or shouts" and "using obscene language" were prohibited, as was anything other than a "moderate use of the horn," which could "cause a scandal." With an eye to public health, the city council also prohibited "transporting cadavers or people with contagious sickness, unless by expressed order of the police or other appropriate authority."[75]

So that motorists completed these steps of registration, inspection, and regulation, the traffic code called for the creation of "a traffic inspector responsible to the municipal treasury who will be charged with ensuring that these rules are followed." The traffic inspector would keep "in the office and in proper order . . . the following books":

> a) a general registry of vehicles, listed successively, specifying their class, category, size, conditions, brand, monthly dues, property owner, with exact address, name of the driver and residence, date of registration and withdrawal, specifying the purpose of the vehicle and if payments are current; b) a general registry of drivers, dedicating a single page to each, in which will be included a drawing of the applicant, registration number, current address, place of birth, parents, age, appearance, with a space left blank for necessary annotations; c) a book of inspections, which will consist of whether the city councilman concurred or not or if the municipal authority authorized the inspection.[76]

The power of the fine would enforce these rules. Excessive speed brought a fine of 5–25 pesos, turning without signaling cost 5–10 pesos, and not having a canopy cover elicited a 5-peso fine, with the potential confiscation of the vehicle by the state if not fixed within eight days.

The Ciudad Juárez rulebook provides a unique window into the rationality of state authorities, or "governmentality,"[77] concerning the regulation of automobiles in the early twentieth century. In Ciudad Juárez, a three-part legal regime controlled automobility, consisting of *registration, inspection,* and *regulation.* Each piece of this regime involved a distinct operation of power, even if these operations were intended to work in concert. In *registration,* personal data was recorded for the purposes of tributary obligations, a fundamental state function.[78] But more than this, registration created a correspondence between car and driver by matching vehicular data to personal data. In *inspection,* tests were administered and judgments made. The aspiring driver's ability to maneuver a vehicle was tested before a three-person jury. The machine, meanwhile, was examined by the traffic inspector and a city councilman to ensure that it was in proper order. Interestingly, inspections concerned more than the fitness of cars and drivers to operate on public roadways. Public order and morality were also considered, as drivers had to have "good morals" and vehicles needed to possess "a clean and decorous appearance." These examinations designed to ensure the safety and morality of the public order evoke Michel Foucault's work on "discipline," which involves both "an observing hierarchy" and "a normalizing judgment" in its subjectification of those it operates upon.[79] The resonance with Foucault's influential work deepens in light of Ciudad Juárez's *regulation* of motorcars. By providing drivers tax books and registration numbers, the latter of which had to be recorded both in the former and on the vehicle in clearly visible script, the state made the legality of the car-driver coupling visible for surveillance by police officers. Through such measures, the power of the law over the automobile took hold.

If the moral aspects of this legal regime seem intrusive, the law also carved out a unique place for the individual. Drivers of private, personal vehicles, for one, were not subject to the same requirements as drivers of public vehicles (who had to carry the rulebook for annotations by the inspector and police and had to get monthly car inspections). It is interesting to consider that in this time of the nascent bureaucratization of the modern Mexican state, drivers of public vehicles would be the ones responsible for purchasing and preserving the documents for noting police officers' observations of them. Drivers of both private and public vehicles also had to carry tax books demonstrating compliance with tributary obligations. Similarly, it was the individual driver who marked the assigned registration number on the vehicle.

Jonathan Simon, remarking on the early administration of automobiles in the United States, observed that "laws governing the operation of vehicles," "civil liability, the general rules of care taking in public life," and "insurance" served to create the responsible driver.[80] The disciplinary regime of automotive governmentality described here—creating records of individual drivers in official records, normalizing behavior through the inspection of driving and property, but also entrusting individuals in their own governance—can also be seen to "responsibilize"[81] the modern car driver in Mexico. And the marked decrease in the rate of traffic fatalities over the last seventy-five years reflects the power of this disciplinary project.

This brief account of the earliest mode of governing automobiles returns us to key themes from the first chapter. To control the collective agency of society, in this case automobility, the state targets materiality—it registers its presence, inspects its function, and regulates its movement. And in governing technology, the state itself is coproduced, not only in the form of new government bureaucracies such as a traffic inspector's office and the police, but also through the collection of taxes to finance the state's growth.

Given the dynamic of state formation through automobile taxation and regulation, it is unsurprising that this regime of automotive governance spread throughout Mexico over the course of the twentieth century. Today, every state in the country possesses a system of automotive administration relying on driver's licenses, vehicle registrations and inspections, and traffic police. In addition, the tax for registering automobiles in Ciudad Juarez eventually morphed into an excise tax assessed for the mere possession of an automobile. Mexican president Gustavo Díaz Ordaz issued a decree in 1964, the Tax for the Possession or Use of Vehicles, to temporarily collect taxes on automobiles to finance the 1968 Olympic Games. Colloquially referred to as *la tenencia,* from the verb *tener* (to have), the tax became an important source of revenue that the federal government ultimately found difficult to do without. *La tenencia* thus became a permanent tax in Mexico, and states adopted similar levies on vehicle ownership. Long subject to scorn among vehicle owners, *la tenencia* was finally repealed by the federal government in 2012, although twenty-seven of thirty-two states still collect it. *La tenencia,* then, provides a clear example of how controlling automobility can contribute to the creation of the state.

3.3.2 *Reducing Environmental Risk*

By the 1970s, the insecurity of automobility began to manifest itself in other ways in Mexico. As the number of cars and duration of daily commutes increased in cities, greater amounts of harmful gases were released into the air, slowly transforming the country's natural environment. Although it is difficult to envision now, given Mexico City's reputation as among the most polluted places on the planet, there was a time not long ago when national writers with no irony described it as "the region with the cleanest air."[a] In the 1940s, visibility in the metropolis extended more than 7 miles. By the 1990s, however, smog, fed largely by automobile exhaust, had reduced it to a little over 1 mile. Not only that, but the paving of roadways impeded rainwater filtration into city aquifers, which aggravated the city's sinking, a structural predicament dating back to the Spanish colonial authority's draining of the lake on which the Mexica had constructed Tenochtitlan. Pavement has also altered the flora and fauna in the city by suffocating the soil and intensifying the sun's rays. In this polluted environment, rates of genetic abnormalities, asthma, conjunctivas, respiratory diseases, digestive problems, and other ailments have increased.[82]

Automobile pollution is representative of the (post)modern condition. As automobility has advanced in Mexico as the primary mode of mobility—driving national economic policy, turning natural resources into valuable commodities fueling development, changing the physical layout of towns and cities, and establishing itself as a cultural reference point—so too has it poisoned the ecological foundations of society to the point that it threatens society's survival. This is the essence of the "risk society," which labors to manage the "hazards and insecurities induced and introduced by modernization itself."[83]

Although federal authorities understood the harmful effects of environmental pollution in the early 1970s, they did not take action to reduce harmful automobile emissions until 1986. A presidential decree passed that year—Measures against Pollution in the ZMCM (Mexico City Metropolitan Zone)—called for the production of unleaded gasoline,

---

a. The phrase *la región más transparente* was popularized as the title of one of Carlos Fuentes's early novels, which focuses on social life in Mexico City. Before Fuentes, Alfonso Reyes, a Mexican poet active in the first half of the twentieth century, also used the expression to describe the city. Reyes, however, took the phrase from Alexander von Humboldt's description of the city in *Visión de Anahuác* (see Fernández Christlieb, *Modernas ruedas de la destrucción*, 34).

restricted the circulation of buses from neighboring states into Mexico City, and established limits for the emission of contaminants (hydrocarbons, carbon monoxide, nitrogen oxide, etc.) in new vehicles.[84] In 1991, the federal government required installation of catalytic converters on new vehicles sold in the country.[85]

Mexico City, meanwhile, the area most affected by automobile pollution, has generally been in the lead in introducing legislation to combat the environmental risk of automobility. Already in the 1970s, the city was conducting vehicle inspections, referred to as *verificación vehicular*. *La verificación vehicular* was made mandatory in 1988. A year later, the municipal government passed the Hoy No Circula (No Driving Today) program, inspired by A Day without Cars, an international movement to raise awareness about the harmful effects of automobiles by having people voluntarily forgo them for a single day (September 22). Hoy No Circula was designed to reduce the number of vehicles circulating in Mexico City by 20 percent; it prohibited each car from operating one day a week, as designated by the last digit of the automobile's license plate number. Thus, for instance, cars with license plate numbers ending in 5 or 6 were prohibited from driving on Mondays, while those with plates ending in 1 or 2 were restricted from circulating on Thursdays.

In 1997, the government combined the two programs—*la verificación vehicular* and Hoy No Circula—in an effort to renovate the city's aging population of cars. *La verificación vehicular* continued as before. However, cars built within the last four years that passed *la verificación vehicular* were exempt from Hoy No Circula restrictions. These cars earned a double-zero (oo) designation, with an accompanying sticker, and were exempt from inspections for two years. Cars built within the last eight years that passed *la verificación vehicular* were also exempt from Hoy No Circula restrictions and received a single-zero (o) sticker. All other vehicles received a two (2) sticker and remained under the Hoy No Circula restrictions. As a result, the original Hoy No Circula goal of limiting 20 percent of the automobile population was reduced to 8 percent.[86] These measures have generally been viewed as a success, despite early criticism that resources to enforce the new provisions were insufficient.[87] Car emissions in Mexico City have been reduced, and air quality has improved considerably.[88] And *la verificación vehicular* has become a model throughout the country, with seventeen states adopting similar measures.

The *verificación vehicular* and Hoy No Circula programs, like the Ciudad Juarez City Council vehicle rulebook, provide insight into the governmental rationality involved in governing the "risk" of automobil-ity.[89] The programs demonstrate how the government's administration of automobility has evolved in concert with its insecurity. To ensure safety on roadways, the first legal regime rested on a triple operation of the registration of the car-driver coupling, inspection of cars and drivers individually, and regulation of moving vehicles. These elements remain in *la verificación vehicular* and Hoy No Circula, but modified to cap-ture and diminish the pollution of automobility.

The purpose of *la verificación vehicular* is the inspection of vehicles. But an important change took place in this program concerning how the government viewed the car. Before, the focus was the operability of a particular vehicle—whether it would function properly on public roadways. Now, the state is concerned with essence of the automo-bile—what the car emits. And although this emission is largely naked to the human eye, the diagnostic equipment used in *la verificación vehicu-lar* realizes a "modification of scale"[90] that renders invisible pollutants visible and measurable and thus subject to the law.[b] In this way, *la verificación vehicular* extends the gaze of the state, through scientific surveillance technology, from the conduct of the automobile (its oper-ability) to its essence (its gaseous emissions).

The inspection in turn creates a registry of vehicles based on pollu-tion levels. Once captured at the inspection center, the "risk," or pollu-tion level, of an automobile is recorded in a registry and "translated"[91] from its scientific terms to a numerical scheme—00, 0, or 2—according to the risk it poses. The vehicle is then entered into the registry of cars permitted to travel freely or not, and a sticker is affixed to communicate the car's risk level to traffic police who, although unable to see vehicle pollution, are nevertheless responsible for enforcing the Hoy No Cir-cula program.

For its part, Hoy No Circula is intended to regulate. But here, too, a change has taken place from prior regulation schemes. Before, authori-ties were primarily concerned with whether drivers were operating their

---

b. The intervention of science and technology evokes the work of Bruno Latour, who in describing the power of the laboratory in Louis Pasteur's battle against the anthrax ba-cillus that was ravaging cattle in France, noted that in the lab, "It [anthrax] is freed from all competitors and so grows exponentially, but, by growing so much, ends up, thanks to Koch's later method, in such large colonies that a clear-cut pattern is made visible to the watchful eye of the scientist" (Latour, "Give Me a Lab and I Will Raise the World," 146). Like Pasteur's laboratory, diagnostic technology exposes the presence of pollutants to authorities' eyes.

vehicles in accordance with the established traffic code. Now, given the "risk" of automobility to society, authorities attempt to prevent drivers from operating their vehicles at all, at least on the days dictated by vehicle performance on pollution tests and placement within the registry.

The *verificación vehicular* and Hoy No Circula programs, then, represent key elements in the regime of risk management that emerged in Mexico during the 1990s to manage the new insecurity of automobility. This regime did not displace the first regime that was based on discipline and making drivers responsible. Rather, it added a layer of legal regulation, as well as the accompanying state agencies required to oversee inspections and monitor vehicles. Interestingly, in contrast to the earlier regime, risk management displays less concern with individual motorists. The moral quality of individual drivers does not concern the state, nor does their ability to operate a vehicle. Rather, it is the materiality of the car that is of interest, both its essence in terms of pollution emissions and its circulation on roadways on prohibited days. And although the invisibility of pollution makes it difficult for law officers to assess the threat of a particular vehicle, the system of numerical stickers applied to inspected cars makes the risk assessment legible for those responsible for monitoring roadways.[c]

3.3.3 Securitizing Legal Uncertainty

Since the 1990s, the risk-management model of governing automobility has spread to other states in Mexico. Seventeen of thirty-two states today operate a version of *la verificación vehicular*. Over this same time, the deepening of Mexican postmodernity—the coterminous processes of economic liberalization, political democratization, and cultural globalization—has revealed new dimensions of automotive insecurity, which are in fact quite old. As first portrayed in *El automóvil gris*, the car today is a threat to public security. Its monetary value makes it a major target of organized crime syndicates.[92] Its mobility is central to the commission of crime, be it drug, arms, or human trafficking or evading police or military forces. Its enclosure of space provides the cover necessary for illicit trafficking. And its combustibility can transform it

---

c. Though the focus of the law has shifted, this new legal regime still affects drivers. Not only must drivers register themselves and their vehicles with the state, but they must also conduct themselves as responsible users of the natural environment, willing to remove their vehicle from roadways or purchase their way into cleaner, less risky vehicles.

into an improvised explosive device for cartels battling one another or the state.[93]

Crime involving automobiles speaks to fundamental gaps in the prevailing system of governing automobility. The registration, inspection, and regulation of vehicles are failing to ensure the security of motorcars. Central to this failure has been the permeability of the state to corruption. The heroic efforts of honest authorities idolized in *El automóvil gris* have not been sufficient to stem the tide of criminality resulting from automobility, as poor compensation[94] and rich opportunities for illicit gains have limited the ability of authorities to be honest, never mind heroic.

In response, governmental authorities at different levels have devised policies to try to regain control of the wheel. In Mexico City, municipal leaders in the early aughts prohibited the use of two-door vehicles as taxi cabs, a measure targeted at the iconic *vocho,* whose two-door design made it a favored vehicle for malevolent taxi drivers preying upon passengers.[95] And at the federal level, the automobile's threat to public security provided the backdrop for President Calderón's launching of the Public Registry of Vehicles.

The legal framework for the REPUVE was actually established on September 1, 2004, when President Vicente Fox—from the same National Action Party (PAN) as Felipe Calderón, and the first opposition candidate to claim the Mexican presidency from the Institutional Revolutionary Party (PRI)—signed the REPUVE law. The Public Registry of Vehicles, as described in the law's first article, "is an information tool of the National System of Public Security, whose goal is to provide public and legal security to those acts undertaken with vehicles."[96] The acts to which the law refers include "the adding or dropping of registrations; obtaining plates; infractions; the loss, robbery, recovery, and destruction of vehicles that are produced, assembled, imported, or circulate in national territory."[97]

But while the legal basis of the REPUVE is recent, its ideological inspiration dates back much further, as evidenced by earlier attempts of the federal government to create a national registry of vehicles. The Federal Registry of Automobiles, established in 1965, was designed mainly to facilitate the collection of taxes and mandate "the registration of vehicles manufactured or assembled in the country, or imported, that were intended for the transportation of people or cargo."[98] A successor, the Federal Registry of Vehicles, was launched in 1977 to bolster the government's ability to collect taxes on vehicles. The registry primarily targeted units from abroad, especially from the United States, a segment of the vehicle population not accounted for in state registries.[d] The registry was eliminated by President Carlos Salinas in 1990, however, ostensibly due to technical difficulties and an inefficient bureaucratic structure.[99] Salinas's decision to terminate the program was nevertheless viewed skeptically and blamed for a subsequent increase in car thefts.[e] It is perhaps not surprising, then, that the next effort by the federal government to control automobility emphasized security.

The National Registry of Vehicles (RENAVE) was signed into law on June 2, 1998, by President Ernesto Zedillo. The new federal vehicle registry sought to "generate a database belonging to the federal government consisting of vehicles manufactured, assembled, imported, or circulating in national territory in order to prevent contraband and automobile thefts."[100] Combining elements from the previous registries, the information stored in the RENAVE was made available to the public and supplied by producers, dealers, insurance providers, and car owners. Distinctly, however, registration in the program required a fee (375 pesos, or 47 US dollars, for new cars). And reflecting the neoliberal age in which it was born, the registry was operated by a private concession, Talsud.[101]

Founded to combat insecurity, the RENAVE fell into disrepute when the head of Talsud, Ricardo Miguel Cavallo, was arrested in Cancún

---

d. The Federal Registry of Vehicles also addressed the economic concerns of both the federal government and national car dealers, who saw imported vehicles as a threat to the industrialization and progress of the nation. The law empowered authorities to "verify the registration and legal standing of vehicles, including being able to confiscate them" (El Secretariado Ejecutivo del Sistema Nacional de Seguridad Pública, *Libro Blanco*, 25). Despite its limited focus on imported vehicles, the registry contributed an innovative public component to vehicle registration, as it allowed people to consult and verify the legal status of their vehicles.

e. The Salinas administration fully embraced neoliberal reforms, such as reducing import restrictions as part of NAFTA negotiations and privatizing various state industries. Reducing the size and scope of government regulation fell very much in line with neoliberal ideology. Today, the consequences of terminating the Federal Registry of Vehicles are largely seen as disastrous. With the termination, "the only authority of the state that served in the identification and registration of vehicles produced, assembled, and legally admitted into the country disappeared." And "[the registry's] disappearance brought with it an increase in crime levels with regard to robberies and the alteration and falsification of documents" (Castro Medina, *Criminalística en la identificación de vehículos automotores*, 20). The Mexican Association of Automobile Dealers (AMDA) concurs, noting that the elimination of the registry left "a legal vacuum" with regard to the administration of automobiles, which resulted in a "shameful increase in the theft of autos" (Asociación Mexicana de Distribuidores de Automotores, *Un siglo en movimiento*, 104).

when he was found to be Miguel Angel Cavallo, an Argentine war criminal wanted by Spanish authorities for torture and other crimes committed during Argentina's military dictatorship in the late 1970s.[102] Among the allegations lodged against Cavallo was that he had enriched himself while working for the Argentine military by seizing the assets of his victims, resources that provided him the capital to begin Talsud.[103] Thus, not only was the head of the program responsible for ensuring the integrity of motor vehicles in Mexico using a false identity, but the head of the program responsible for protecting the property of Mexicans had enriched himself by seizing the property of people he had tortured.

Leaving aside the more dramatic personal elements of the story, the RENAVE evidences the growing concern among government leaders with vehicular security at the turn of the last century, a concern at the center of the REPUVE. And the REPUVE, in fundamental ways, attempts to learn the lessons of the RENAVE's failure in order to construct a more successful federal registry of automobiles. First, the REPUVE does not have an enrollment fee, as the RENAVE did. Second, the RENAVE proved that the "operation of the Vehicular Registry under the scheme of a public service concession presented insurmountable complications that made its performance unviable."[104] As a result, the REPUVE functions as a public program, under the authority of the Executive Secretariat of the National System for Public Security (SESNSP).[105] Finally, while the REPUVE preserves the RENAVE's concern with insecurity, it exhibits a particular formulation of security that reveals its governmentality. In the REPUVE, the link between registration and public security is indirect. In a white paper describing the registry, the SESNSP explains, "The fact that the REPUVE is an instrument of the National System for Public Security does not mean that it is targeted to the search, pursuit, localization, or recovery of stolen vehicles in the country." Rather, "the aim of the Public Registry of Vehicles is to provide public and legal security to the acts that are undertaken with vehicles."[106]

"Public and legal security," in the eyes of the SESNP, has been compromised by a host of factors. Primary among them has been a "disproportionate growth of the vehicular fleet circulating in the country," around 6 percent, or 610,000 vehicles per year. This growth presents unique challenges to the government, "since at the same time that the automobiles increase, so too do the legal acts or facts involving their participation," including "production, importation, sales, rent, financing, insurance, theft, repair or destruction, among others. . . . As a consequence of the increase in the number of vehicles, the number of activities in which they were involved grew, exceeding the capacity of the government to offer certainty to citizens regarding the identification and the legal circumstances of every automotive unit."[107] This, combined with "the intense commercial traffic that exists along the Mexican borders, principally with the United States of America . . . require[s] our country to have, for security's sake, a registry system capable of detecting the origin and destination of all the vehicles circulating on national terrain."[108]

Of course, a number of automobile registries already exist in Mexico, especially at the state level, that identify vehicles and provide legal status. These are the registries that arose under the first legal regime described earlier in this chapter. However, the SESNSP explains,

> such instruments do not satisfy the necessity that the citizenry has that the federal government guarantee the legal certainty *[certeza jurídica]* with respect to the legal situation of vehicles. Because the registries were developed by multiple legislatures, they are not similar in the type of data they contain, nor do they possess the certainty that the physical characteristics of the automobiles correspond to the information about them. In addition, they possess a multitude of aims, as some find themselves targeting fiscal concerns rather than providing certainty. In addition, the data contained in the different registries were not made to be consulted either by federal authorities or the citizenry.[109]

Interestingly, then, it is not a lack of regulation or absence of authority that breeds the insecurity that the REPUVE aims to respond to. Rather, it is the multiplicity of regulations, registries, and authorities governing automobility that diminishes the "legal certainty" of vehicles.

The concept of legal certainty occupies a central place in the SESNSP's white paper. "National security is strengthened," the document explains, "by providing public and legal certainty to the acts involving automobiles, since in consolidating and publicizing a reliable and effective vehicular registry, anonymity is suppressed and the commission of criminal acts is inhibited."[110] Further, "the REPUVE was imagined as an answer to the government's pressing need of providing legal certainty to the presence, situation, and existence of all the vehicles moving in Mexican territory. Such certainty strengthens national public security directly, since to the extent that the behavior of the automotive registry is fully known, it will be possible to control and diminish the incidence of crimes committed against a vehicle or with a vehicle."[111] Thus, "although the REPUVE is not directly charged to combat car thefts or related offenses, in offering certainty and making transparent the legal situation of automobiles, it is clear that it inhibits criminal activity and deters the commission of illicit acts."[112]

The SESNSP's description of legal certainty resonates with key ideas introduced in the last chapter. In noting that as a consequence of "the increase in the number of vehicles, the number of activities in which they were involved grew, exceeding the capacity of the government to offer certainty to citizens regarding the identification and the legal circumstances of every automotive unit," the SESNSP speaks to how the increasing materiality of mobility in Mexico is reducing the state's ability to govern. And in explaining that the existing "registries were developed by multiple legislatures" and "are not similar in the type of data they contain," the SESNSP attests to the problem of redundancy and heterogeneity of state agencies dedicated to the administration of collective agencies such as automobility. The consequence of increasing material agency and redundant state agencies is an "uncertainty" about the things through which collective agency happens.

To ensure the legal certainty of automobiles, the law arms the REPUVE with two types of advanced information technologies: (1) a database integrating the data of every vehicle in the country, and (2) radio-frequency identification (RFID) tags that adhere to vehicles to transmit their identifying data as they circulate. The SESNSP describes the database as "the prime material of the REPUVE." It is "a multidimensional solution of variables and values that stores in an ordered manner and in a grand electronic archive the data on vehicles moving through national territory."[113] The information in the database includes "vehicle identification number," "the essential characteristics of the vehicle," "the name and home address of the property owner," and "information provided by federal authorities and federal states in accordance with this law,"[114] which includes "the inscription/registration number assigned by the SESNSP."[115]

Data are provided to the database from three distinct types of entities with reporting obligations to the REPUVE: *autoridades federales* (federal authorities), including agencies such as the Secretariat of Finance and Public Credit, which manages customs in Mexico, and the Secretariat of Communications and Transportation, which is charged with the administration of federal highways; *entidades federativas* (federative entities), the state-level authorities such as departments of motor vehicles and justice departments that manage vehicle registries in their jurisdictions; and *sujetos obligados* (obligated subjects or liable parties), the private-sector businesses dealing with automobiles, such as

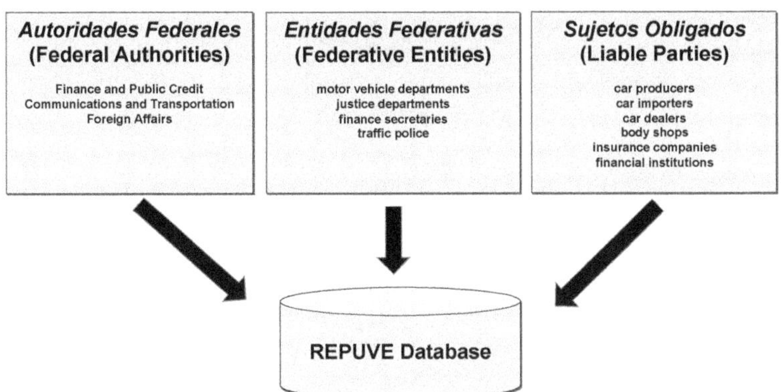

FIGURE 20. Conceptual design of the REPUVE database.

manufacturers, importers, financing agencies, and insurance companies (fig. 20).

The SESNSP, as the authority in charge of the registry, oversees a General Directorate of the Public Registry of Vehicles consisting of four subdirectorates, each responsible for a different area of the program. The State and Federal Operations Implementation Directorate, for instance, is charged with supervising the program's adoption by the *entidades federativas* and *autoridades federales*. The Relations with Obligated Subjects Directorate, meanwhile, ensures compliance with the REPUVE law via the *sujetos obligados*. The Procedures and Citizenry Directorate is responsible for managing the public's contact with the program. And the Oversight and Verification Directorate is responsible for technical aspects of vehicle inspections (fig. 21).

As the organizational structure of the REPUVE indicates, the SESNSP does not simply administer the flow of data into and out of the registry's database; it also ensures compliance with the REPUVE law. Thus, the law empowers the SESNSP to conduct "ordinary and special verification visits to the *sujetos obligados,* who must permit access and provide information that the personnel require for the fulfillment of their work."[116] *Sujetos obligados,* further, are subject to fines for any of the following infractions: "enrolling a vehicle in the registry after the deadlines established in these procedures"; "not enrolling a vehicle in the registry"; "not presenting reports referred to in the law"; "making unauthorized use of the proofs of registration, documents, and other means of identification related to the registration of vehicles," "altering, omitting, copying, or permitting illicit registrations or notices, regis-

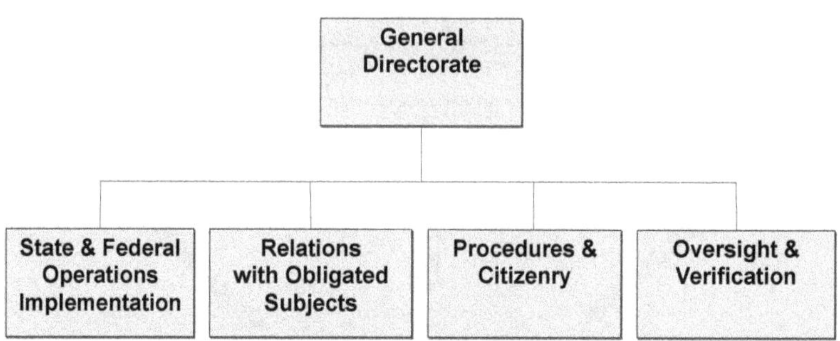

FIGURE 21. Organizational structure of the REPUVE.

tering false data, providing false information or information to unauthorized users or third parties, providing information from the Registry without authorization, or not reporting irregularities as required"; and "making use of the information, documents, or receipts of the Registry for profit, whether directly or via a third party."[117]

While the REPUVE law places the SESNSP in charge of administering and enforcing the rules of the registry, the database itself, "in treating a topic of national security," is stored in the Plataforma México system that operates out of the federal police bunker (see chapter 1). Information entered into the registry by law remains there permanently, even if the car exits the country, is destroyed, or cancels its registration at the local level.[118]

Access to the database, meanwhile, is broad, which is in line with the government's conception of "legal certainty." As the SESNSP notes, "The information in the REPUVE database acquires its true value when it is consulted and utilized."[119] To this end, the REPUVE law dictates that the "*autoridades federales* and *entidades federativas* that provide information from their registries to the [REPUVE] registry will have access to the information contained in it." In addition, "any citizen or public entity can inform themselves, without charge, of the legal situation of a vehicle in Mexican territory," although the registry is prohibited from providing personal data, except if one is the owner of the vehicle in question or has been authorized to access such information.[120] The REPUVE provides access to its database through three separate mechanisms: a call center; email consultations; and a webpage, where individuals can access information about a particular vehicle by entering its plate number, vehicle identification number, or registration number.[121] Once accessed, the registry provides the following information: "make, model, year, vehicle class, type, registration number, plates, number of doors, country of origin, engine displacement, number of cylinders, number of axles, and legal status of vehicle."[122]

These data are detailed but not personal. Thus, a person consulting the REPUVE database can know whether the vehicle she has purchased or is interested in acquiring has been stolen or altered. In other words, the inquiring party would have "legal certainty" that the vehicle is a genuine and lawful object.

The dataset, however, is only the first technical component by which the REPUVE aims to increase the legal certainty of automobility in Mexico. The more exotic element is proof of registration *(constancia de inscripción)*, referred to among REPUVE administrators as *el chip. Los chips* are actually RFID tags or stickers 4.13 inches wide by 2.76 inches high (fig. 7), adhered to vehicles once they have been entered into the registry. At the center of each sticker lies an integrated circuit or microchip, based on ISO/IEC 18000–6C RFID standards, which can store 800 bits of information and transmit that data via radio frequency. The RFID tags, chosen by the SESNSP after technical consultations with the National Autonomous University of Mexico (UNAM), the National Polytechnic Institute, and the Institute of Technology and Higher Education of Monterrey, and produced by the Neology corporation, are passive, which means they transmit data only upon being activated by a RFID reader. In addition to data transmission via RFID, each tag possesses other security elements: a holographic texture in order to prevent its falsification, a miniform number and barcode in its lower portion, microtext containing its batch number and the inscription "Registro Público Vehicular México," and an imprint of the national coat of arms in invisible ink. In addition, a tag becomes inoperable when detached from the windshield, making its transfer impossible.[123] Each chip has a life span of ten years, "during which its physical characteristics remain the same without affecting its functioning. In addition, it is able to withstand different climates and continuous exposure to ultraviolet rays."[124]

The RFID tags arrive from the Neology corporation attached to a piece of paper, or "miniform," and vehicle data are inscribed on them through proprietary software.[125] The only information entered onto a tag is the vehicle's VIN. "The stickers are not applied by the General Directorship of the Public Registry of Vehicles, but are distributed among the *dependencias federales, entidades federativas,* and *sujetos obligados,* who apply them."[126] *Sujetos obligados* who produce or import vehicles simply record the VIN onto the chips, which they adhere to vehicles, and then record the link between the VIN, chip number, and originating miniform number in the database. Laser printers then print out the mini-forms and tags.[127] In the case of cars already circulating in Mexico, the RFID tags are applied by *autoridades federales* or *entidades federativas*. In such cases, a "physical inspection of the vehicle as well as its documents" must first be conducted. A tag is only printed and applied "when trained inspectors corroborate that the data of the vehicle correspond with that in the documents, and no other illicit modifications have been made" to the vehicle.[128] Once a tag is printed and applied, "the last phase of the process is to verify the correct functioning of the RFID." To do this, the vehicle is passed under "a fixed radio-frequency reading portal," whose radio signal determines whether the data on the RFID sticker can be read.[129] When the data are verified, the enrollment process is complete.

The laws and documents describing the Public Registry of Vehicles, like the *verificación vehicular* and Hoy No Circula programs and the Ciudad Juárez vehicle rulebook before them, provide insight into the governmental rationality informing the state's effort to overcome the legal uncertainty of the automobile in contemporary Mexico. The basic logic of registration, inspection, and regulation remains. But the changes to these operations under the REPUVE illuminate how governmental power shifts through the adoption of advanced information and surveillance technologies.

Beginning with registration, the REPUVE does not seek to merely create a registry of car-driver couplings, as the first disciplinary regime of automobile governance sought, or a registry of vehicle emissions, as the risk-management model did, or a registry of any particular aspect of automobility. Instead, it endeavors to create a registry of registries, to merge the different records kept by *entidades federativas, autoridades federales,* and *sujetos obligados* into a single database that the SESNSP manages and that is stored on Plataforma México.[130] This "registry of registries" mirrors a central trend noted by surveillance scholars toward the "convergence and integration of different surveillance systems."[131,f] However, in pushing for the systematic integration of data from the *entidades federativas* and *autoridades federales,* the federal government demonstrates not only an interest in data merging but also a lack of trust in the administrative capacities of the state agencies entrusted

f. In Kevin Haggerty and Richard Ericson's "surveillant assemblage," surveillance today is being "driven by the desire to bring systems together, to combine practices and technologies and integrate them into a larger whole" (Haggerty and Ericson, "Surveillant Assemblage," 610).

with the management of automobility. In essence, the state agencies that were "co-produced"[132] with automobility in order to govern it are now seen as obstacles to automobility's governance. And by creating a mechanism for the systematic review and integration of their data, the REPUVE provides the federal government a way to "police the police" or "govern the governors."

Turning to inspection, the REPUVE preserves the practice of systematically examining automobiles. But what is examined has changed. The proper conduct of the machine is not what is judged to ensure the safety of the driver and others. The gaseous emissions of the machine are not what are diagnosed to preserve the natural environment. Rather, the body of the vehicle is inspected to certify its identity. To guarantee vehicle identity, the REPUVE inspection requires technicians to locate three instances of corresponding VINs. This use of the vehicle body to verify identity is a key theme in surveillance studies.[g] Biometric technologies attest to this trend, for example fingerprint and palmprint scans, facial recognition, iris scans, and gene mapping. Of course, in the case of the REPUVE, the body that interests the state is not human, but machine; and the markings it attempts to document are not those left by nature or biology, but by industrial manufacturers in accordance with international standards for identifying vehicles.

The regulation of automobility is also altered under the REPUVE. Previously, state agencies were created or charged with watching over the conduct of vehicles or their circulation during times of prohibition. Now, RFID tags and readers and license plate recognition technology are deployed to verify both the identity of vehicles (by matching the information stored on RFID tags to license plates and records stored in the REPUVE database) and their location (by noting the place of the RFID and license plate recognition readers). The state here displays an interest in the "presence" of vehicles.[133] This concern with tracking movement resonates with the work of surveillance scholars too, who have identified a trend toward "enforced locatability" in the electronic monitoring of "dangerous persons" such as criminal offenders through RFID and GPS technologies that "produce a sense of human proximity without the element of physical presence."[134]

g. "In a world of identity politics and risk management," notes David Lyon, "surveillance is turning decisively to the body as a 'document' for identification, and as a source of data for prediction" (Lyon, *Surveillance Society*, 72.)

This "enforced locatability" is present in the REPUVE, but its technological composition alters the operation. On the one hand, since RFID tags are affixed directly to vehicles, this regulation is not "participant dependent,"[135] requiring the active participation of people enrolled in the program. Indeed, once applied, *el chip* automatically communicates its information whenever it passes a corresponding reader, absolving the individual driver of any obligation to check in with authorities and denying her of any knowledge that she has been checked in with authorities. On the other hand, in recording the "presence" of vehicles automatically, the RFID tags and readers, license plate recognition technology, and digital databases of the REPUVE eliminate the need for police officers to regulate automobility. The surveillance technologies of the REPUVE in this sense "displace"[136] the human agents of the state who have historically been responsible for supervising automobility. Through these mechanisms, the regulation of automobility becomes more automated.

In sum, the structure of power at work in the Public Registry of Vehicles remains the same as in previous regimes of automobile governmentality. The automobile continues to be governed by a process of registration, inspection, and regulation. But the introduction of surveillance technologies alters the operation of the process (fig. 22). And these alterations, in turn, reveal much about the evolution of state power in contemporary Mexico.

3.4 PROHESION: THE WAY TO MAKE THINGS STICK

Various terms have been used to describe the increasing presence and impact of surveillance technologies on contemporary society. Roger Clarke in the late 1980s developed the term "dataveillance" to describe how surveillance had moved from a physical monitoring of individuals to "the systematic use of personal data systems in the investigation or monitoring of the actions or communications of one or more persons."[137] Mark Poster, also early in the information-technology revolution, modified Foucault's concept of the "panopticon"[138] to fit the digital present, coining "superpanopticon" to describe how "circuits of communication" and the "databases they generate" constitute "a system of surveillance without walls, windows, towers or guards."[139] Gary T. Marx, who has studied surveillance for decades, uses the term "new surveillance" to distinguish how monitoring is now done "through the use of technical means to extract or create personal or group data,

HISTORICAL ERA	REGISTRATION	INSPECTION	REGULATION
Governing Road Safety (1910s-present)	Vehicles and Persons	Capability of Drivers & Operability of Vehicles	Compliance with Traffic Rules
Governing Pollution "Risk" (1980s-present)	Vehicles and their Emissions	Essence of Vehicles	Abstinence
Governing Legal Certainty (2010s-present)	Vehicles and their Registries	Materiality of Vehicles & Integrity of Records	Localization

FIGURE 22. Governing automobility in Mexico.

whether from individuals or contexts."[140] Kevin Haggerty and Richard Ericson modify the Deleuzian concept of "assemblage" to name the "surveillant assemblage," which "operates by abstracting human bodies from their territorial settings, and separating them into a series of discrete flows [that] . . . are then reassembled in different locations as discrete and virtual 'data doubles.'"[141] David Lyon uses the term "surveillance society" to highlight the ubiquity of surveillance technologies in everyday life and to stress, contrary to dystopian trends in other works on surveillance, that "surveillance—watching over—both enables and constrains, involves care and control."[142]

The governmental rationality on display in the REPUVE shares many points of connection with these ideas. The registry monitors automobility by processing data rather than watching over their circulation. This monitoring occurs without the policing agents that defined previous modes of surveillance—it is accomplished instead through technical means that record the identity and location of vehicles at a particular time and place and transmit the data to another. And it is designed to provide both care and control, as ordinary Mexicans can access the information in order to have greater confidence in buying and selling vehicles, and the security forces can use the same information to track stolen vehicles. In this sense, the REPUVE embodies "dataveillance," the "superpanopticon," "new surveillance," "surveillant assemblage," or "the surveillance society."

At the same time, it is difficult to apply these concepts wholesale to Mexico. For one, the concepts are often cast widely, broadly describing the applications, significance, and consequences of surveillance technologies on society, including not only how relations between governors and the governed change but also how intimacy, family, and recreation are affected. The REPUVE, however, responds to a particular problem of governance—the uncertainty of automobility—that cannot be assumed to have a significance beyond the state.

Nevertheless, these concepts do speak to a changing relationship between the state and citizens, whereby the former utilizes surveillance technologies to keep the latter under regular monitoring. But in the REPUVE, the particular problem that the state is grappling with does not concern human subjects, but the state and the automobile. There are too many databases consisting of too many assorted formats unable to communicate with one another, and there are too many vehicles involving too many diverse legal acts for the government to keep track of. What concerns the state here is not human subjectivity, but the state agencies and material agency that underlie the exercise of automobility in society.

To respond to these challenges, the REPUVE endeavors not to track individual subjects or even human populations, but to integrate the disparate databases tracking automobility in Mexico and to attach RFID tags to vehicles to establish their "presence." This is a double operation, involving processes distinct from those usually associated with surveillance. And it thus invites conceptual work that might capture its novelty.

In searching for the words to describe the operations of power in the REPUVE, the image of "sticking" keeps coming to mind. It is an image inspired by the RFID tags used in the program. With these tags, the state adheres itself through its technological delegates to the material substance of automobility. And in doing this, the state aims to make automobility more adherent to the law. Thus, adhesion is a dominant theme. The image of sticking also applies to integrating the databases of the *entidades federativas, autoridades federales* and *sujetos obligados.* That is, the state is interested in having its different agencies become more consistent and unified, or cohesive. Cohesion, then, is another theme. Adhesion and cohesion. The root in both words is the same, *haerere,* a Latin verb that means "to stick."[143] And it provides a simple way to imagine the double operation of power present in the REPUVE. To control automobility, the state attempts to make the state agencies and private actors governing automobility more cohesive with one another and to make the materiality of automobility more adhesive to the infrastructure of the state.

These reflections suggest that collective agency in society more generally possesses a *hesion,* or viscosity, that can be made either more or less amenable to governance independent of individual subjectivity. The amenability of automobility to governance, for instance, can be manipulated through the automobile itself—by maintaining records of a vehicle (manufacture, sale, registration) that create its history, by placing markings on it such as vehicle identification numbers or license plates that identify it to authorities, by attaching an external surveillance technology such as an RFID tag that allows it to be tracked by authorities, or by designing it with certain features such as black boxes that allow it to be tracked in time and place. Or it can be influenced through the material environment through which it circulates—by installing obstacles such as toll booths, by altering the surfaces of roadways with rumble strips, or by modifying the number of traffic lanes. Or it can be affected through the organization of state agencies that police it—by changing the arrangement of authority both within departments (chains of command) and between them (organizational structure), by revising the composition of departments through staffing choices, by keeping records on vehicles, and so forth. The REPUVE, then, can be understood as an effort to increase the hesion of automobility in Mexico by using surveillance technologies to alter the composition of automobiles (by attaching RFID stickers on them), their physical environment (by installing RFID scanners on roadways), and their policing (by integrating registries).

Speaking of the state's adoption of surveillance technologies in terms of hesion rather than surveillance enables us to imagine governance outside the purely human drama of governors and the governed. As a quality inherent to collective agency, hesion directs our attention to the other elements besides humans that enable activity in society. This is not to say that people are not involved. Politicians pass the legislation that establishes the rules governing collective agency, and citizens or individuals are responsible for ensuring compliance with the law. But the concept pushes us to imagine dynamics of social control outside human subjects and inside the material artifacts and state infrastructures through which collective agency happens.

Introducing *hesion* to describe the Mexican government's novel program to govern the uncertainty of automobility in Mexico is not to suggest that the phenomenon is new. State authorities since the founding of Mexico have sought to make agency more tractable by manipulating its conditions. In the case of mobility, this was done during colonial times by erecting *presidios,* or forts, near roadways to secure commercial transports from Chichimeca and bandit raids. In the case of communications, the material culture of the indigenous peoples of Mexico was destroyed and schools were established so that the symbolic basis of Spanish culture would take root in the conquered region. In the case of identification, differences in the physical appearance of peoples inhabiting colonial Mexico were transferred into *pinturas de casta* that instructed viewers on the racial hierarchies being imposed on society.

So the state's practice of manipulating the hesion of collective agencies in Mexico is not new. But at the same time, the REPUVE, as well as the other efforts of the Mexican government to enroll surveillance technologies to fight crime, does reveal a certain postmodern condition, or post-postmodern condition, which makes *prohesion,* which I define as governmental efforts to increase the *hesion* of collective agencies, especially significant. First, the materiality of collective agency today is increasing and/or changing. In the context of economic liberalization and global production, the number of cars, mobile phones, computers, and so forth in Mexico today is increasing. As a result, the vibrancy of mobility and communications is enhanced and becomes more difficult to govern.

Second, the existing infrastructure of the state has become less capable of governing agency. The pace of change in mobility, communications, and identification has left behind the state organizations that were "co-produced" over the course of modernity to govern collective agencies. The neoliberal policies of the Mexican government were intended to diminish the state's oversight of automobility. The Salinas administration terminated the Federal Registry of Vehicles in 1990, and the Zedillo administration entrusted the RENAVE to a private firm. And corruption remains a pervasive problem for the state in Mexico, which reduces its capacity to enforce the law.

Third, the alternatives available to the Mexican state to address the insecurities of collective agency are limited by the course of history. If, before, the state could control the number of automobiles in Mexico through restrictions on production, importation, and pricing and the use of telephone communications through the monopolization of telephone services, its embrace of neoliberal political economy emphasizing free trade, smaller government, and international standards of democratic governance restricts its ability to manage society as it had done in the past.

Fourth, surveillance technologies provide state authorities with novel tools that allow it to reimagine the art of governing in a way unparalleled perhaps since the dawn of statistics.[144] If, prior, the ordering and control of society was accomplished through shaping human subjectivity through disciplinary techniques[145] or manipulating populations through control tactics,[146] computer systems, RFID, GPS, and biometric technologies now allow states to govern collective agencies without having to engage unreliable human subjects or depend solely on corruptible human agents. The materiality of collective life can be directly engaged instead. Thus, while *hesion* is not new, the challenges of governance that have accrued over Mexico's recent past and the technological tools laid before state authorities today make *prohesion* a more fitting form of governmentality than discipline.

It is not surprising, then, that when the Mexican government has had to grapple with the challenges of mobile telephony and identification in its War on Crime, it has developed programs—the National Registry of Mobile Telephone Users (RENAUT) and the Citizen Identity Card (CEDI)—that also feature prohesive strategies. As verbal communication has migrated from telephone wires to radio waves, the state has struggled to get a handle on communications, especially in cases involving kidnappings. In response, the RENAUT creates a governmental database of cell phone lines and their subscribers and requires cell service providers to create their own databases storing subscribers' names, addresses, fingerprints, and photographs and to provide geolocalization of individual calls,[147] all of which exhibits a push for the *cohesion* of records and data-processing procedures involving mobile telephony. The RENAUT also mandates cell users to register their phones with the registry by sending a text message with their name and date of birth, or Unique Population Registry Code (CURP),[148] a move akin to *adhering* state RFID stickers on automobiles. Thus, an activated phone, like a passing car, can be identified and localized immediately. The CEDI provides a single form of identification to replace the innumerable forms (birth certificate, driver's license, voting card, advanced electronic signature for tax payments, military service card, etc.) that Mexicans otherwise contend with, a *cohesive* idea. And the CEDI bases personal identification on biometrics such as photographs and iris scans that are embedded into a government-issued card,[149] which evokes having the body *adhered* to the state's governmental infrastructure.

In the REPUVE, RENAUT, and CEDI, then, one witnesses the emergence of a distinct mode of governmentality for the control of collective agency. Whether the brave generals, inspectors, and police officers depicted in *El automóvil gris* ever approximated their historical counterparts and inspired their successors, the emergence of *prohesion* speaks to their place in history. At the dawn of the twenty-first century, insecurity would not be overcome by having men of the law bring bandits before firing squads, but by deputizing advanced information and surveillance technologies to make the things of mobility, communication, and identification stick.

CHAPTER 4

Ni con goma

"If you go up two floors, no one in this building even knows
what the REPUVE is. They know that we're here on this
floor. But they don't know what the program is . . . "

"Until they have their car stolen. Then they know what it is."

—Lucas Espinoza and Daniela Flores, REPUVE administrators

4.1 TEN THOUSAND CALDERÓNS

On April 10, 2010, an important deadline loomed for mobile telephone
users in Mexico. By this date, cell users needed to register their phone
lines with the government's National Registry of Mobile Telephone
Users (RENAUT) or risk termination of service.[1] Users could register in
one of two ways: by sending a text message to the number 2877 with
the word "ALTA" (Spanish for "subscribe"), together with their name
and date of birth or their Unique Population Registry Code (CURP); or
by going to a service provider to have their data recorded.[2] Registrations at service centers would also record user fingerprints as a security
measure.

The deadline did not pass without controversy, however. Public reaction to the RENAUT was highly critical. The registry was said to violate
the privacy granted by Article 16 of the Constitution, which sets out
people's right to not be disturbed in their home, property, or documents
except upon a court order.[3] And various lawsuits challenged the legality
of the registry.[4] The critical reception extended to service providers as
well. They refused to comply with the RENAUT's requirements to collect biometric data. And they vacillated on the question of suspending
service to subscribers who did not register their phone lines. Movistar,
which controls 20 percent of Mexico's cellular market, announced that
it would not suspend service; Telcel, which holds majority market share

and is owned by sometimes "world's richest person" Carlos Slim, stated that it would suspend lines but allow continued text service.[5]

Fittingly, given the ethereal substance of mobile communications, the controversy extended into the digital realm. In the days and weeks before the deadline, the Twitter page #RENAUT exploded with messages critical of the program:

> For my part (and I believe many others), #RENAUT is not the solution and I do not agree with it, let's say #NOalReanut [sic] -> http://ow.ly/10e2Y

> I say #No to Big Brother! Campaign against dystopia. http://www. noalgranhermano.org/ #acta #renaut #mexico

> Remember the scandal about the national registry of autos? Do you trust registering your cellular phone? #RENAUT #mx

> "Possible FAIL for #RENAUT . . . we can register ourselves with fake info and more than once . . . hahaha!! http://www.milenio.com/node/403397 Give it a RT

> They Can Take Away Our Cell Phone Numbers But They Cant Take Away Our Freeeeeeeeeedooooooooom #renaut http://tinyurl.com/yzvn585" [original in English]

> Did you know the #RENAUT plans to include your fingerprints in a second stage? Check out: http://bit.ly/ajTyTW #BigBrotherFail

Twitter messages such as these not only conveyed people's antipathy toward the telephone registry but also included key details about its operation that readers might not already know. The registry would include fingerprints. And more importantly, one could register multiple times using false data.

Various webpages offered additional information on the registry's operation. One blog, *Trucos para evadir RENAUT* (Tricks for evading RENAUT), featured an image of a naked foot with its sole facing the reader and its middle toe extended in simulation of a middle finger. It gave readers strategies for duping the registry, for example, what the blog called the "zombie" technique: "We search for deceased Mexicans on the Internet or in a newspaper. We copy the data of the deceased and register them. That way we have a live phone, but with calls from the great beyond." Another option was the "chameleon," targeting social networks such as Facebook and HI5: "We search, in whatever network, user profiles that display names, dates of birth, and places of residence, preferably those with advertising attached to them, since they are usually already registered with their CURP." The post goes on, "There are also various social networks of professionals (accountants, lawyers, engineers, etc.) where users display their full name and date of birth. Also, we can casually add friends to Facebook or HI5 to learn their data. Once we have recorded the data, we can go the [federal government's] CURP Web portal and fill out the profile data for the users, hit click, and BINGO! it shows you their CURP, which we will send to RENAUT, who accepts it gladly. This is a bit tedious, but completely reliable. 100% tested. This is how I registered."[6]

The "chameleon" strategy highlights an interesting characteristic of Mexico's Unique Population Registry Code. The algorithm that generates the CURP is publicly available, making it possible to calculate other people's codes, so long as their full name and date of birth are known. On this basis, those critical of the RENAUT pushed the envelope further, detailing how a simple Wikipedia entry for someone like President Felipe Calderón could be used, in combination with the National Population Registry's CURP portal, to determine his CURP and register one's phone line with it.[7]

When the April deadline passed, 70 percent of the 83,500,000 mobile lines in Mexico were registered with the RENAUT. This left 25,202,935 lines at risk of disconnection. However, the digital campaign against the RENAUT left its mark. Newspapers reported that upward of 50,000 of the millions of telephone numbers registered were fictitious, with 10,000 registrations made in the name of President Felipe Calderón.[8]

But the National Registry of Mobile Telephone Users was not alone in its difficulties. The Citizen Identity Card (CEDI) was mired in its own problems. The identity card had been announced by Calderón in July 2009. Already by August 2009, the newspaper *La Jornada* published opinions of officials from the Federal Electoral Institute (IFE) who criticized the CEDI on the basis that Mexicans would confuse it with their voter card. The voter card, according to the officials, had served as the de facto form of national identity for the last two decades and the government had invested $4 billion in it. Alberto Alonso y Coria, executive director of the Federal Electoral Registry, explained that the voter registry contained the biometric data of seventy-eight million people of voting age and that the voter card should "survive and remain under whichever scheme, because it is an instrument that not only has a broad social acceptance, but also the legal capacity to identity Mexican citizens and provide security in the most important event for the IFE, voting. The ID allows us to guarantee that only those who have the right to vote do so and that they only do so once."[9] In the eyes of the IFE, the Citizen Identity Card was not only a waste of taxpayer money but also a threat to the integrity of democratic elections in Mexico, a right only recently won. And as one of the most respected institutions in Mexico, after the Catholic Church and the military,[10] the IFE's opposition to the CEDI helped engender a wider opposition to the card and placed the fate of the program in doubt.

The Public Registry of Vehicles (REPUVE), meanwhile, was encountering its own complications. Unlike the other two programs, these were born not of notoriety but of a lack of familiarity altogether. When I visited Mexico City in summer 2010 to collect data on the REPUVE, I learned that no one in my immediate circle of friends and acquaintances had heard of the program. Indeed, no one I spoke to around the city knew of it, despite the pomp and circumstance accompanying its launch the previous summer. Invariably, in their effort to be polite to a misguided *gringo,* people explained that what I was describing sounded much like the mobile telephone registry, which was regularly in the news that spring. The *defeños'* (residents of Mexico City) ignorance of the automobile registry was understandable. Although the program aimed to register, inspect, and regulate the nearly twenty-five million vehicles circulating in the country by 2012,[11] only a handful of Mexico's thirty-two states (San Luis Potosí, Veracruz, Baja California Sur, Colima, Sonora, and Zacatecas) were applying radio-frequency identification (RFID) tags to vehicles in summer 2010, thus leaving the future of the registry in considerable doubt.

These vignettes demonstrate that if Mexico's mobile telephone registry, identity card, and automobile registry evidence a distinct mode of governmentality, "making thing stick" is more easily done in theory than in practice. State strategies for gaining new holds on communications, personal identification, and mobility meet with multiple obstacles. People refuse to comply with measures they deem invasive. Companies balk at the financial costs associated with new programs. And politicians, public officials, and state governments oppose these efforts for political gain and to protect their own domains of power. As a result, the viability of these programs is left in question. Thus, while the basic logic of the RENAUT, CEDI, and REPUVE can be described as *prohesion,* the diverse points of resistance that the programs encounter give the impression that "nothing works," a sense captured by the Spanish phrase *ni con goma,* which is used when things "do not fit" or "go together" but literally means "not even with glue," an apt phrase to describe the failings of programs intended to "make things stick."

This chapter concentrates on the resistance that state surveillance in Mexico engenders. While the term "resistance" might immediately bring to mind concerned individuals organizing themselves to oppose the government, the stories that make up this chapter offer a more expansive view of defiance in sociolegal contexts. Over the past three decades, scholars have come to conceive of resistance in terms of the everyday "weapons of the weak"[12] that ordinary people draw on to oppose the "common place of law"[13] in their lives. The uneven histories of the RENAUT, CEDI and REPUVE, however, demonstrate that major points of resistance to state projects derive from other sources as well, including political and bureaucratic structures formed over the course of modern history, cultural formations residing at the core of national histories, and technological and design elements embedded in monitoring programs. As the state looks to gain a greater hold over collective agencies in society that are distributed across a diverse collection of actors, institutions, and material arrangements, it encounters resistance all along that distribution. To understand the power, or lack thereof, of surveillance technologies in contemporary society, then, one must not only consider their design, nor just individuals' reactions to them, but the range of resistance they encounter.

4.2 RESISTANCE IN THEORY

Resistance is part and parcel of any effort to exercise control. Michel Foucault made this point bluntly with his oft-quoted dictum that "where there is power, there is resistance."[14] It is James Scott, however, who provides the definitive treatment of resistance in his influential work *Weapons of the Weak*. Based on research with Malay peasants, Scott claims that the weak do not counter authority through the "peasant rebellions," "peasant revolutions," or other forms of organized resistance that have usually captured the imagination of scholars. Rather, they do so through "everyday forms of peasant resistance," which consist of "foot dragging, dissimulation, false compliance, pilfering, feigned ignorance, slander, arson, sabotage, and so forth." In contrast to organized resistance, such as strikes, marches, or political campaigns, these mundane forms of resistance "require little or no coordination or planning; they often represent a form of individual self-help; and they typically avoid any direct symbolic confrontation with authority or with elite norms."[15]

This is not to say that resistance cannot take organized forms. Although Scott does not engage it directly, the literature on social movements and collective action describes tools that subjects can employ, in addition to the "weapons of the weak," against those in authority.

Whether arising from the rational choices of participants[16] and mobilization of resources by organizers,[17] political opportunities at a given historical moment,[18] the "historicity" of contemporary societies,[19] the construction and maintenance of collective identity,[20] the strategic framing strategies of organizers to the broader public,[21] or the emotions and meanings imbued in issues and events,[22] collective action is able to produce the force necessary to both resist the debilitating effects of power and remake the social landscape by effecting political change,[23] legislative action,[24] or—even when failing to achieve stated goals—raising public consciousness around key issues.[25]

But *Weapons of the Weak* highlights the effects that seemingly innocuous "everyday forms of resistance," devoid of the lofty goals of collective action, can have. "When such acts are rare and isolated, they are of little interest," Scott explains, given their inability to affect the operations of hierarchical relations in society. "But when they become a consistent pattern (even though uncoordinated, let alone organized) we are dealing with resistance," which counts as "any act(s) by member(s) of a subordinate class that is or are intended either to mitigate or deny claims (for example, rents, taxes, prestige) made on that class by superordinate classes (for example, landlords, large farmers, the state) or to advance its own claims (for example, work, charity, respect) vis-à-vis those subordinate classes."[26] "The intrinsic nature and, in one sense, the 'beauty' of much peasant resistance," he adds, "is that it often confers immediate and concrete advantages, while at the same time denying resources to the appropriating classes, and that it requires little or no manifest organization."[27]

The withholding of resources to the appropriating classes is vital for understanding the true force behind the "weapons of the weak." To make the point, Scott offers the example of peasant desertions from the Russian Army in 1917 during the height of the First World War.[28] The desertions were not motivated by high-minded aspirations to topple the czarist regime and fight for a progressive political agenda. The poor souls who found themselves conscripted into the Imperial Russian Army simply wanted to return home to care for their families. But these self-interested acts, coalescing into a wave of desertions, led to the collapse of the czar's main institution of repression and hastened the fall of Nicholas II and the Russian imperial order.

Scott's work on peasant resistance has profoundly influenced the social sciences. In the case of sociolegal research, this perspective helped pave the way for understanding the "common place of law." In this view, the law is not a mere tool of authority that instantaneously endows those possessing it with a mystical power to dictate relations in society. It is instead a process that can provide the subaltern unanticipated opportunities to resist and counter those in positions of authority. This is so whether one speaks of a Filipino peasant in the US territory of Hawai'i, whose refusal to have his stigmata treated disrupts the formation of identities central to the exercise of power in the occupied land, or of the feminist movement's activism to change domestic violence law to protect women from gender violence.[29] In the "common place of law," people can oppose authority through "resistance against law," which means refusing to comply with its dictates; "resistance by means of law," which is using the legal system to challenge authority and privilege; and "resistance which redefines the meaning of law," by adopting the law's language and legitimacy to challenge established notions of justice and the social world.[30]

This literature on resistance provides a good starting point for understanding the difficulties experienced by the National Registry of Mobile Telephone Users, the Citizen Identity Card, and the Public Registry of Vehicles. And the next sections borrow from it to describe the positions that the public and business community assumed against the surveillance programs. However, the obstacles faced by the federal government in implementing its security programs extend beyond these traditional sources of resistance. The surveillance technologies in Mexico's War on Crime have confronted four distinct types of challenges: the concerns of citizens, the misgivings of the business community, the composition of the state and interplay of political interests within it, and technical constraints. Because each is meaningful to the outcomes experienced by the Calderón administration in attempting to reform the state's security apparatus, the following review encourages a broader understanding of resistance in sociopolitical contexts.

4.3 THE OPPOSITION OF ORDINARY MEXICANS: PRIVACY, SECURITY, AND DISTRUST

The digital campaign against the RENAUT reveals that many people in Mexico took a clear position "against the law." These individuals were critical of the law. And they sought to sabotage its operation. But what would bring ordinary people in Mexico to stand against the RENAUT? This section examines three interrelated sources of uneasiness: concerns over privacy, insecurity about the security of the state, and mistrust of the state forged over time.

4.3.1 *Privacy Concerns*

Over spring and summer 2011, I distributed a survey about the RENAUT, CEDI, and REPUVE and security in Mexico to ninety-eight individuals from upper-middle and working-class neighborhoods in Mexico City and rural Zacatecas. The survey briefly described the programs and asked respondents to answer a series of questions about them, including "Do you support this program?" and "Why?" When responding to the "Why?" question, respondents were prompted to provide a short answer.

In the case of the RENAUT, 52 of 97 (53.6 percent) respondents answered "No" to the question of supporting the program. The most common reason identified in their short answers had to do with "privacy":

To me, [this program] should not have been implemented because it
   controls people and violates privacy.
I don't agree [with the program], there is already no privacy.
They take away our privacy but you can't do anything about it
   because it's mandatory.

Thus, people in Mexico, like people in other countries asked about their opinions on government surveillance,[31] felt that having to register their phone lines would result in a loss of privacy.

4.3.2 *Security Insecurities*

Interestingly, however, privacy concerns in Mexico involve more than unease about what the government might know about you. As frequently, people expressed the following concerns:

Personal data are used for other purposes.
I believe that the information is sold to criminals.
There is so much corruption and you don't know who to trust.
It [the registry] won't help; if the data are used, they'll be used by
   criminals inside of the program.
Many times they sell information to other companies.

As these statements reveal, while Mexicans are concerned about the government violating their privacy, they are as concerned about criminal elements or other third parties gaining access to their data. In other words, they are not so much worried about the "right to be let alone," as US Supreme Court justice Louis Brandeis famously described the right to privacy, as they are about the security of their personal data.

There is a touch of irony to Mexicans' concerns about data security, since the telephone registry is designed to bolster security. More importantly, these concerns increased skepticism about the program, which dampened the number of people registering their phones ahead of the April 10, 2010, deadline. But these doubts were not exclusive to the RENAUT. In the case of the Citizen Identity Card, this apprehension about privacy and, more specifically, data security was repeated. Asked to give their opinions in support or opposition of the CEDI, 52 of 97 (53.6 percent) again responded "No." As to "why," those surveyed noted, "It's bad that now the government wants to keep us under watch," "We already have IFE, why do we need more?" and "Because they will always be checking in on us." But data security again was uppermost in their minds: "There's no guarantee that my data will be in good hands," "Because I no longer trust the president," and "I'm not sure it's legitimate; it doesn't give me confidence."

Believing that the government works hand in hand with criminal elements or is willing to sell citizens' private data to the highest bidder might sound overly suspicious. But it is worth remembering that these opinions harken back to historical events in Mexico that remain present in people's memory. The Twitter user at the beginning of the chapter who remarked, "Remember the scandal about the national registry of autos? Do you trust registering your cellular phone?" provides a prime example of this. The registry the user references is the National Registry of Vehicles (RENAVE), the notorious for-profit program operated by Ricardo Cavallo, the Argentine war criminal, who did indeed use the registry's databases to target vehicles to steal. In this case, the criminal was very much a part of the government. Following that sordid tale, in 2002, all of the data of sixty million Mexican voters contained in the Federal Electoral Registry and managed by the Federal Electoral Institute, the main opponent to the new identity card, was sold by the Mexican company Soluciones Mercadologías, which had been contracted to manage the database, to the US company Choice Point.[32] Choice Point in turn resold the data, for the price of $1 million per year, to the US federal government, whose Border Patrol used it to identify Mexican migrants crossing the Mexican-US border. Thus, data entrusted to the government has been sold to third parties and used against Mexican citizens before. These stories remain present in the popular imagination of Mexicans and are a ready well of mistrust when they encounter programs such as the mobile phone registry, national identity card, or automobile registry.

Politicians and those working with the programs openly acknowledged the challenge these ghosts of Mexico's administrative past posed in the present. As Mexican senator Miguel Angel Chico Herrera expressed when discussing the CEDI in a 2011 newspaper interview, "The population has lost trust in authorities and it has to be said openly that these identity documents could end up on the black market, as happened with other documents . . . the registry of vehicles or the voter roll that the IFE has, and so many other documents, identifications, or lists of citizens that have appeared on the black market."[33] Diego Avila, meanwhile, a technician working at the REPUVE site in Zacatecas that I visited, noted much the same. "There are 20 percent of the people that simply are not going to come, they're not going to come [to register]. Maybe they have bad information. They think that this is insecure, that we work for criminals, like RENAVE. RENAVE was a bad program that is distorting the REPUVE, because people think that it's the same story." This left those responsible for implementing the program, such as Samuel Gallo, an official in the REPUVE's State and Federal Operations Implementation Directorate, in a position where they had to, as he explained it, "combat the history of RENAVE."

4.3.3 Government Mistrust

At the heart of Mexicans' historical memory and their critical comments about the three security programs is a mistrust of government, a view of the government as corrupt, illegitimate, and/or ineffective. At times, these views are expressed explicitly. In certain media outlets, for instance, the Citizen Identity Card has been deemed of dubious legitimacy because the amendment to the Population Law calling for its creation "was passed in 1990 . . . as a mere expression of the will of the head of the executive branch"—Carlos Salinas—who has gone down as the most unpopular leader in modern Mexican history after Porfirio Díaz.[34] Implemented before Mexican democracy returned, the CEDI is seen to lack the weight of the law.

Similarly, on blogs and opinion pages, people doubted that the government would ever be able to enforce the sanctions associated with noncompliance. Commenting on the RENAUT, for example, Ernesto Villanueva, a law professor at the National Autonomous University of

Mexico specializing in information rights, opined that "it's going to be impossible to eliminate [people's cellular] service or impose a sanction on 40% or 60% of cellular users because they will never have a cell phone in their name. Please! We live in Mexico. If the most basic elements of public security cannot be guaranteed, how are they going to selectively apply the law to those Mexicans and the national cellular industry?"[35] Because the state is unable to control crime in the first place, it is hard for some observers to believe that the state will be capable of policing something as sophisticated as mobile technology or of contending with the influence of cell service providers.

The historical memory of Mexicans thus bears on their experiences and expectations of surveillance programs in the present. As James Scott mentions in describing the roots of resistance beyond self-interest, many forms of resistance

> may be individual actions, but this is not to say that they are uncoordinated. . . . It is, for example, no exaggeration to say that much of the folk culture of the peasant "little tradition" amounts to a legitimation, or even a celebration, of precisely the kinds of evasive and cunning forms of resistance I have examined. . . . In this and in other ways (for example, tales of bandits, peasant heroes, religious myths) the peasant subculture helps to underwrite dissimulation, poaching, theft, tax evasion, avoidance of conscription, and so on. While folk culture is not coordination in the formal sense, it often achieves a "climate of opinion" which, in other more institutionalized societies, would require a public relations campaign.[36]

In Mexico, the bizarre stories surrounding previous efforts to register people and things enter into popular culture and serve as touchstones that work against similar programs in the present, placing many Mexicans in a position "against the law" of prohesion.

4.4 THE CONCERNS OF COMPANIES: COSTS, CORRUPTION, AND CONFUSION

Ordinary Mexicans—whose mobile phones, multiple forms of identification, and automobiles are the focus of the surveillance programs—were not the only ones opposing the programs. Companies and businesses responsible for the manufacture, sales, and service of the technological artifacts at the heart of mobile telephony and automobility also resisted the move toward prohesion. As noted at the start of the chapter, cellular service providers varied in their support of the RENAUT. Movistar publicly opposed the program, stating that it would neither provide users' data, collect biometrics, nor cut services to those who did not sign up. Automobile producers and importers also expressed reservations about the REPUVE. And like the survey respondents, history loomed large.

In an interview with Tomás Ayala and Vicente Bautista of the Procedures and Citizenry Directorate at the REPUVE, Bautista explained that the RENAVE was casting a shadow over how companies were approaching the REPUVE. "The companies are providing information, let's call it confidential, about the company as well as the dealers and the purchaser," he explained. "The management of this information is a concern because, yes, there was a problem before. The system failed. Who knows what happened with the information that was there. . . . So obviously, the industry asked us, 'Whose hands are we putting this information in?' . . . They didn't accept the REPUVE blindly." But if companies shared ordinary Mexicans' concerns about privacy and data security, the main obstacles for the business sector in general were the implied costs of complying with the new surveillance programs, the corruption perceived in government operations, and the confusion surrounding the programs.

4.4.1 *Business Costs*

Unsurprisingly, perhaps, the operational costs of complying with the new surveillance programs figured most centrally in many companies' thinking. In publicizing their reservations about the RENAUT, the mobile telephone industry estimated that complying with the registry law would cost it $100 million.[37] Such costs were also a primary sticking point for the automobile industry concerning the REPUVE. Fernando Orozco, a representative with the Velocity Motor Corporation (VMC), a major car manufacturer in Mexico, explained to me, "The truth is that we as a car producer never wanted the REPUVE. Why? Because it generates a cost for us. It generates a very high cost that cannot be reflected in the [price of the] vehicle. It's a severe operational expense." Agustín Sandoval, a representative from Sucaro, a car manufacturer importing vehicles to Mexico from Asia, sounded a similar note of critique when I interviewed him: "The responsibility [for tagging vehicles] is ours, not the state or city. Now, politically speaking, we didn't want to accept the responsibility for adhering the chip. We said, 'Let the government do it!,' 'Put up modules [for chips] where plates are given out!,' 'Let the government do it!' Why us?"

The operational and administrative costs are clearer to see the closer one gets to the production line. VMC provided me a tour of its plant outside Mexico City. Company representatives noted with clear pride the efficiency of their production process, explaining that a vehicle comes off the production line every eighteen seconds. It was precisely this precision that the REPUVE tags threatened. "We cannot stop the flow of vehicles," Orozco explained as we toured the facility. "So we had to adapt ourselves to have the flow be the same, so that it would be continuous and would not stop, whether we're talking about a sedan or a pickup." But at the same time, the REPUVE law required car makers to ensure that cars coming off the line had a REPUVE sticker on them. "A vehicle doesn't leave the plant if it doesn't go through the REPUVE. If a vehicle doesn't have a sticker, or a registration with REPUVE, we cannot sell it," Orozco said, "and there we are running a risk because we are risking a sale. And that hurts business."

Fitting this operation into production lines with a global reach was a challenge. "At this site, we have two assembly plants," Orozco continued.

> Plant 1 produces sedans. Plant 2 produces trucks. But then we have a third shipping point for what we call different market characteristics. [That third shipping point] is where we have vehicles going to the United States, Brazil, and Europe. That line doesn't have the REPUVE because they are exports. . . . Periodically, however, we receive domestic vehicles here. So this means having the [REPUVE] infrastructure in three spots. . . . Like I said, this doesn't add any value to the vehicle from the business point of view. . . . Whatever addition that is going to be added to a vehicle has to be controlled under various measures of quality. All of this has an expense, a logistics. . . . We return to the same point. This [RFID tag] didn't have a benefit for us as a company.

The direct costs of applying the tags did not exhaust auto producers' friction with the program. Aesthetics also mattered. This point was first shared with me by Tomás Ayala and Vicente Bautista at the REPUVE's Procedures and Citizenry Directorate. "The sticker," Ayala plainly said, "is not the most decorative element." As a result, he continued, "the car producers put up a lot of obstacles for sticking on the tag." The VMC representatives I met with confirmed these points of discord. "That a vehicle that is sportier, and more expensive then, has to have this sticker," Fernando Orozco opined, "for us that is horrible."

Beyond aesthetics, the industry was also concerned about safety and the impact that the tags might have on drivers' visibility. While VMC and Sucaro were placing the stickers on the upper-middle part of windshields, above the rearview mirror, the ideal placement recommended by the REPUVE, they harbored concerns. Orozco at VMC explained, "I'm not going to lie that the best option economically and technically would be to place the chip on the lower right-hand side of the windshield, because you as a driver in Mexico, all of the driving is on the left. It wouldn't obstruct a traffic light or a sign on that side. Some manufacturers put it on the upper-left side. But putting it on the lower-right side would have practically no consequence. That is, it's there visually, but there's no risk to visibility."

4.4.2 *Corruption Concerns*

Transnational corporations that produce and import vehicles in Mexico were not the only companies concerned about the Public Registry of Vehicles. Following my field visit to Zacatecas, a local newspaper ran a blurb about my research. Once the piece appeared online, I was contacted by Jonathan Vargas, a representative from the SecureRead company, which had competed for the contract to produce REPUVE tags. That concession was ultimately awarded to the Neology corporation, a company owned by Alejandro Burillo Azcarraga, a member of the influential Azcarraga family, which founded the Televisa television corporation. Burillo himself has a wide variety of business interests, ranging from soccer clubs to mobile telephone providers. On the phone, Vargas told me that he felt the need to share his version of the story of how Neology won the concession for REPUVE tags.

The concession, as national newspapers reported, was based on a trial of different RFID technology providers that took place at the Autódromo Hermanos Rodríguez, a well-known racing circuit in Mexico City that hosts Formula One races. The competition was presided over by technical advisers from three of Mexico's leading universities: UNAM, the National Polytechnic Institute, and the Institute of Technology and Higher Education of Monterrey.

Describing his company's experiences with the competition, Vargas started by focusing on the differences between active tags, which SecureRead specialized in, and passive tags, which Neology produces:

> The operational minimums that they showed in the presentation can correspond only to an active tag. They had to be able to read multiple vehicles in traffic, be able to read from a mobile terminal that was in a patrol car, and be able to read from the side of the road at a certain speed. And I think they

wanted 99 percent accuracy. All those are things the active tags can do, not something the passive tags can do. . . . We thought this presentation matched something that we can deal with. So, we wanted to do the trial.

"At the Autódromo Hermanos Rodríguez," he remembered,

we were the last to go. We were one of two active-tag companies. Everybody else was passive. . . . So we were the last ones. We went immediately after a company called Apex. They were having some problems with the readers. They were having problems with passing the bridge. And other things like dirt and rain. Anyway, at the end of the day it started raining and pouring. And the guys were like, "You are not going to be able to work." And I said, "No, absolutely we can work. We're active. It's different." The professors that were evaluating the technology, they knew nothing about active. All their protocols were all about passive tests, you know one vehicle, one per lane, not in traffic. . . . They knew active existed, but they hadn't designed a protocol for it. . . . Anyway, we run our test and then we had a representative from one of the other large companies come up to us [after] and say, "Hey, why are you here?" We said, "Well, you know we are running this test." And he said, "No, sorry, it's already been decided. Neology is going to win. That's what I've been told." At first, I was a little shocked. So we asked for a meeting with this Campa Cifrián, with the head [of the Executive Secretariat of the National System for Public Security (SESNSP)], the guy that was running REPUVE, and effectively that's what we were told . . . "It's has already been decided, it's going to be passive. I can't tell you who is going to win, but it is going to be passive, you are too expensive."

Vargas's account carries a tinge of sour grapes. But it should be noted that corruption in public procurements in Mexico is common and problematic. Indeed, the Organisation for Economic Co-operation and Development has remarked that the lack of a competitive bidding system encourages collusion between bidders and reduces the viability of international suppliers.[38] What is more, SecureRead was not the only company flummoxed by the conduct of the trial. After its completion, the REPUVE competition was featured in media reports that cited a "lack of transparency in the results of the evaluation decision."[39] Firms that took part, including Integra, Mobil-link, and e-Plate, complained that they did not have sufficient details on the types of tests to be done prior to the competition.[40] Roberto Campa Cifrián, head of the SESNSP mentioned by Vargas, was also accused of being tied to one of the companies with a stake in the bid, the Cosmocolor Company, which produces holograms for stickers like those used in the REPUVE and is owned by Jorge Kahwagi, whose father is president of Mexico's Confederation of the National Chambers of Commerce.[41]

In addition to the intrigue about Campa Cifrián's links to companies bidding on the REPUVE contract, Vargas's story references a central technical issue at stake in the competition—passive versus active RFID technology. Experts cited in media reports argued that the passive tags chosen by the SESNSP "have disadvantages such as their read coverage and the cost of some of the reading equipment (like arches at highway exits), which go from 2,000 to 25,000 dollars." Active tags, meanwhile, "cost from 12 to 25 dollars [and] have readers that cost from 300 to 500 dollars, some as small as cellular phones." Put another way, "With the active technology, the reader identifies all the tags located in a determined area, which facilitates their search in wide spaces, while the reading of the passive tag is individual and at specific points."[42] Given the seeming advantage of active tags over passive tags, the media reports added to the controversy surrounding the bidding process.[a]

By the time I was conducting my research, the national directors of the REPUVE within the SESNSP considered the controversy over active versus passive tags "a debate already decided" and one that "owed to different interests" rather than technological considerations. Tomás Ayala and Vicente Bautista shared Campa Cifrián's view that while "the active tag had advantages from the point of view of coverage, . . . the two points that were of greatest importance for us were the transferability of the tag . . . and the fact that the battery at that time could last, at best, four or five years." Ayala stressed, "The moment that these batteries stop functioning, the tags would obviously stop functioning, and you would have to devise a replacement of the batteries and all that that would mean for twenty-five million vehicles." But if the program administrators considered the matter closed, the debate, even if

a. In the midst of the controversy, Campa Cifrián was called to testify before the Internal Revenue and Public Credit Commission of the House of Deputies, Mexico's lower legislative chamber, to explain his decision on the bid. He had earlier explained that the decision was based on recommendations of the evaluation team from the three universities, who took into consideration the life span of active versus passive chips, the lower costs per unit for passive chips, and the fact that passive technology was open source and not subject to proprietary restrictions. Unconvinced by the his explanation, the committee demanded Campa Cifrián's resignation and cancellation of the concession, because passive technology was insufficient to stop organized crime and vehicle theft (de la Luz González, "Eligen chip inútil contra robacoches"). The passive technology stayed. But a week later, Campa Cifrián did resign (de la Luz González, "Roberto Campa renuncia al SNSP"), although questions persisted as to whether his boss, Gerardo García Luna, the powerful head of the Public Security Secretariat (SSP), had penned his letter of resignation (Rodríguez, "Campa Cifrián—García Luna").

fed by disgruntled businesses and headline-seeking reporters, had cast the REPUVE's bidding process in doubt and, in the process, diminished the program's legitimacy in the public eye.

4.4.3 Program Confusion

In addition to car producers and tag manufacturers, car dealers make up another another *sujeto obligado* (obligated subject) from the business sector that have a role in implementing the REPUVE. Dealers are a key actor for the registry because, in selling vehicles to the public with chips already adhered to them, they serve as a point of contact between the registry and the public. In summer 2012, I visited dealerships around Mexico City to get a sense of their role as informal ambassadors for the Public Registry of Vehicles. When I asked salespeople and other employees, "What are the tags on the vehicles for?" answers varied. Most described them as "customs stickers" applied when "when cars are imported, so this [sticker] is for when vehicles pass through customs."

These responses are partly true. Mexican customs offices are required to apply REPUVE stickers to vehicles that are imported directly from the United States. And the customs database feeds into the REPUVE database. But the responses also fundamentally confuse the purpose and operation of the REPUVE, which is to provide for "legal certainty" of vehicles by applying a tag to every vehicle in the country without distinction of country of origin. Of the ten dealerships I visited, only one offered a description of the tags that matched the REPUVE. "It is a government rule that [the sticker] has to be there," the salesperson explained, "because if it isn't, the car can be considered stolen, that it didn't enter the country legally."

When I mentioned these interactions in an interview with officials at the Relations with Obligated Subjects Directorate, Fernando Nava acknowledged that "there is a large lack of knowledge present, above all concerning the composition of the REPUVE for the *sujetos obligados*. Windshield installers or body shops are *sujetos obligados* by law who protect and maintain the tags. However, many windshield installers, I don't know if they are aware of the program or if they aren't dealing with the REPUVE. They do what they want with the chips." Given this lack of information, Nava continued, more had to be done to "inform dealers why the chips are there, so that they don't remove the stickers without knowing why they're there, and to inform buyers in case they ask, so that they know how to answer them. [There could be] cases where a new vehicle is damaged and the windshield had to be changed, and the vehicle doesn't have a tag then. Then, if buyers were to come in and see the vehicle and don't see the tag, they would say, 'I want my tag.' Maybe they don't know exactly what it is, but they want it."

But if the car dealers did not know why the stickers appeared on vehicles, they were unequivocal on the question of whether they could be removed. Every dealer I spoke with said they could not be removed. But rather than referring to the REPUVE law by name, respondents noted that, "on the streets, traffic police check stickers often, to see whether the car has it or not. And if they're going to stop you and you don't have the sticker . . . " Or, as another salesperson told me, if your vehicle does not have the sticker, "then later you can have problems if one day you are stopped or you have your car stolen or you have an accident and they are going to check the numbers in the registry, and if you don't have it, there's going to be trouble."

This is an interesting point to reflect upon. The dealers do not understand why the REPUVE tags are on their vehicles, but they are unwavering in communicating that they cannot be removed. In this sense, to put it in sociolegal terms, the dealers stand "before the law" and follow its dictates without knowing what the law actually is.[43] This dissonance can be explained, perhaps, by the history of policing automobility in Mexico more generally, which was covered in the last chapter. The presence of regulatory stickers on vehicles is nothing new in Mexico. And in Mexico City, pollution controls are enforced in part through stickers, which the police monitor. Thus, if the REPUVE sticker is itself is an unknown entity, it represents something common to those working around automobiles and thus fits into a broader pattern of experiences and practices surrounding that thing.

The experiences of car dealers with the REPUVE also contrast in interesting ways with those of car producers. If car dealers were ignorant of the REPUVE law, car producers were only too aware of it, and they actively disliked it. But unlike mobile phone users refusing to register their phones, manufacturers did not stand "against law." Why the difference? Quite simply, the risk of sanctions for noncompliance with the federal law was too great. Julieta Salazar, from the REPUVE's Relations with Obligated Subjects Directorate, told me that "the private sector has always been very strict, above all the big companies, with relation to legal compliance with the federal and state governments. It's not easy for their legal teams to know about a law or regulation and not comply with it."

This point was repeated by the company representatives I spoke with. But additional factors besides the power of the law came into play. "Why does VMC comply with the REPUVE law?" Fernando Orozco reflected. "VMC basically has 100 percent compliance with the law, not only the REPUVE law, but with all the laws, economic, import, customs, et cetera. We are the most precise and maybe the most compliant in the entire industry. . . . [The REPUVE] is a severe operational expense. However, the law was passed. The regulation was made. We had to comply." Here Orozco finishes by noting the lack of options available to VMC to ignore the REPUVE law. But additionally, he speaks with a certain satisfaction about the compliance rate of VMC with the law in general. Legal compliance, to him, is an extension of his company's exacting approach to industrial production that allows it to send a vehicle off the production line every eighteen seconds. Thus, a certain company culture was present at VMC that leaned toward compliance.

In the case of Sucaro, national culture was cited as central to compliance. Sucaro is based in Asia. And Agustín Sandoval and Felipe Ibarra felt this helped explain their company's stance with the REPUVE. "We, as a subsidiary company, report absolutely everything to Asia," Sandoval told me. "A program such as this has so many implications, legal as well as social, from the point of view of drivers. We translate all of this, including the rules and regulations, so that it is absolutely clear. On other hand, the Asian philosophy is very precise, very dignified in the sense that if the law asks me to do this, I have to comply." Ibarra added,

> There are companies that are more rebellious, that in a given moment would sue and not comply. But the Asian companies aren't like that. Without mentioning brands within the automotive industry, there have been two extremes, from two powerful companies. One takes any minor pretext to say, "I'm done with the *chip*. Let the government do it if it wants. I quit! No more!" The other has said, "I already invested millions into the manufacturing process, from the wheels on down, everything is almost robotized. The process of putting on *chips*? I can do it, I already spoke with corporate about it.

These responses seem to suggest that VMC and Sucaro had assumed a stance "before the law," choosing to comply with its dictates, whether because of the exacting nature of the company or its national corporate culture. But this does not give the entire picture. In an email, Julieta Salazar explained that the auto producers "accept it [REPUVE], but are insistent" that the government take over the job of applying tags to vehicles. "It's always a point of discussion," Salazar continued. "In fact, right now, there is a lawsuit through the Mexican Association for the Automotive Industry (AMIA) to reform the law where they propose that the states do it. . . . They insisted that the federal Secretariat of Public Security (SSP) would be the better option for conducting this activity. Even when there is an agreement with the Association, they haven't stopped in their initiative to reform the Public Registry of Vehicles Law so that that obligation would be done away with." In this sense, the automobile industry can be seen to assume a position of "resistance by means of law."[44] Companies like VMC and Sucaro have complied with the REPUVE, analyzing and altering their production operations in order to apply tags to vehicles without disrupting car sales. But at the same time, through their national association, they are using the legal avenues available to them to eliminate the legal requirements that they view as especially burdensome.

4.5 THE OBSTRUCTIONS OF STATE: RESOURCES, LEADERSHIP, POLITICS, AND ORGANIZATION

If the stances of laypeople and business actors toward the RENAUT, CEDI, and REPUVE can be mapped onto existing frameworks for comprehending resistance, this correspondence begins to unravel when considering another key point of opposition to the programs: the state itself. As noted at the beginning of the chapter, the primary source of opposition to the Citizen Identity Card came from the Federal Electoral Institute, an independent governmental body charged with regulating elections. In the case of the National Registry of Mobile Telephone Users, strong public opposition enabled politicians to question the program, and the Communications Commission of the House of Deputies called the RENAUT's leadership to testify about the program's difficulties.[45] With the Public Registry of Vehicles, meanwhile, states have been reluctant to begin applying tags to vehicles.

This is a critical point. One might assume that the state would support government initiatives, especially in the realm of security. But as Daniela Flores with the REPUVE's State and Federal Operations division described to me, there are multiple reasons why working with the state proves difficult: "One is resources, two is the will to do things, three is the political aspect." To this list could be added the organization of state power.

4.5.1 Budgetary Constraints

A federated state's implementation of a program such as the REPUVE requires a sizeable investment in resources. These include, at a bare minimum, computers to process information, printers to print out tags, handheld RFID readers to activate and verify chips once they are adhered, and facilities and salaries for workers in the program. Some of these resources can be paid for by the federal government's Support Funds for Public Security (FASP), monies provided to the states for public security expenses. But discrepancies between allocations and costs can be great. As one member of a Mexico City delegation explained at a national meeting on the REPUVE that I attended, "We have a registered vehicular roll of four million vehicles, we received three million tags, and from the federal project budget we have 7.5 million pesos assigned this year. But we have estimated that the implementation will cost 240 million pesos."

But tags are only a portion of the program costs. To fully register the presence of a vehicle at a transit point, the REPUVE also requires license plate recognition (LPR) technology. LPR technology, however, "has the disadvantage of being very expensive equipment," a point emphasized to me by Ignacio Meza, the official with REPUVE Zacatecas. Workers there also believed that the program needed better facilities than the trailers they were operating out of. Matías Luna, a data specialist, thought that "for this program to function better, adequate space is needed":

> That includes a tarp. That includes a space for the clients, for when, say, four, five, six vehicles arrive. Normally, at least two people come in a car. If five vehicles come, you have ten people—where do you put them? Sometimes the sun is very strong, and twenty minutes standing out here in the sun is a lot. It burns. So I think that there should be an adequate space for the clients to wait for the fifteen to twenty minutes that the process takes. For the technicians too. They do the physical work. It's hard. So, the same thing. Have a tarp or a tent to protect them from the sun, so that they can do their work despite inclemency of the weather.

Another member of the REPUVE Zacatecas team felt that more resources were needed to attract people to the program: "The program, in terms of its goal of installing chips, has certain deficiencies. Advertising, for example. They haven't been able to advertise it in a way that the people in the state understand its purpose or importance for why vehicles will have a chip. This is the reason why there have been few people coming for the installation of chips."

Without the dedicated funds for the program, many states announced suspension of its implementation. For example, Fernando Manrique Rivas, director of Vehicular Control in Morelos, explained that the REPUVE "isn't only putting on a sticker, it's putting in place all of the infrastructure that is going to read and keep the institutions of public security informed. So it is a costly program that today doesn't have sufficient resources. For that reason, the program is suspended."[46] One detects an element of political opportunism in such actions. States shutting down or threatening to shutter federal programs for lack of resources may simply be maneuvering to secure more resources from Mexico City, a dynamic reflective of the strong regionalism that has long defined the country.[47] Indeed, this angle was noted by the State and Federal Operations Implementation members I spoke with. Lucas Espinoza explained, "There is also the political part. The federal government can be pressured to give money that is required. And if San Luis [Potosí] gets more, then Zacatecas is complaining to get more too. Because at the end of the day, it's not their money, it's the citizens' money."

There is a prudential element at work here as well, which is potentially troubling for a program like the REPUVE. Because the FASP used by the states to implement the program are not earmarked and are intended to cover all state public security projects, the more invested in the REPUVE, the less goes into other security priorities. As Ignacio Meza from REPUVE Zacatecas noted,

> There are states that have, apart from the federal resources, invested money from their own governments, against their own accounts. But they are few. Two or three states have done that, like Veracruz and San Luis Potosí. They're achieving the same numbers [of registrations] as we are. But other states that are taking money from the public security funds, they are putting money into REPUVE and leaving little for public security. In the case of Zacatecas, public security is being given priority, and this leaves small change or few funds for the REPUVE. From the same funds, you have police training, equipment, and other things.

Not only that, Meza continued, but "if I invest in an advertising campaign to tell people what the program is about, I don't have the technical capacity to respond." That is, an advertising campaign might help attract drivers, but the resources invested there would come at the expense of program capacity. He added, "The moment that I do the advertising campaign, I am going to have a rush for fifteen or twenty days in which the whole wide world, even if they don't have cars, are going to go to the webpage and want a consultation. This would overwhelm the system, it would be a disaster. We are at a crossroads of technology, necessities, and the rest, which doesn't permit us to advance at the pace that we would have wanted." Thus, a cycle of lethargy befell the REPUVE. States were unwilling to invest in a program they did not see as a priority and that lacked popular demand. But if that demand were there, they would not be able to handle it. So the most rational course of action was to not invest in the program at all.

This situation bred frustration among the data specialists and technicians working at REPUVE Zacatecas. Part of the frustration was understandably selfish. Matías Luna confided to me, "A pay increase would be good. I don't know how much it should be, but I would like it if it were more. Really, whatever you earn, people make what they make, but it's never enough." But the frustration of workers involved more than personal interests. "One of the things that I don't think has functioned well is the interest of the government to start things off well," Diego Avila, a technician in Zacatecas, told me. "San Luis has taken off. Veracruz has taken off. They have taken off because of the interest of the government. They have put everything into it. Here, it's like they're a bit afraid, or they've forgotten about us."

The frustration extended beyond the frontline workers in the program. Ignacio Meza faulted the fact that the amount of the FASP received by states was based on population rather than expediency. "Take a state like Chihuahua," he told me, "which doesn't have the population that the Federal District [Mexico City] has, but its crime level is extremely high. It doesn't receive the funds that it should. The Federal District receives more, although the crime level isn't so great. So there are questions about the formula [for allocating resources]. At the end of the day, the resources are there."

Further, the frustration over resources extended to the national leadership of the REPUVE as well. "Another obstacle that we have encountered," explained Samuel Gallo with the State and Federal Operations directorate, is that "we don't even have the support here to have the budget that would allow us to push certain things forward." Elaborating, he said, "There are different factors that prevent the process from being more agile. A first point is resources. The program doesn't have an adequate communications system. There are not sufficient blueprints. There is not sufficient equipment. It's a significant obstacle to our being able to share information."

4.5.2 Political Leadership

This frustration over resources speaks, in turn, to the second aspect Daniela Flores saw as lacking in the federal government's implementation of the REPUVE: the will or leadership to get things done. "Look," Samuel Gallo lamented, "this project has a very large potential, but sadly it has not been seen as such in the upper spheres, not at the presidential level, nor the interior secretary level, nor the Public Security Secretariat level. They haven't given the program the impulse that it should have." He continued, "It's the political question, it's the question sometimes of leadership, in a way it's a lack of leadership to shake things up." These comments echo the exchange between Lucas Espinoza and Daniela Flores that begin this chapter. Even within their own building, the offices of the SESNSP, the REPUVE team felt invisible. Not only their work, but the importance of their program was being ignored.

The lack of will to get things done extended down from the federal government to the states. As Gallo explained, state authorities, such as the attorneys general in charge of tracking vehicles, were often remiss in updating their databases: "There are national accords that the prosecutors, and even governors in some cases, signed and promised to send information in no more than seventy-two hours in the case of stolen vehicles. . . . However, for questions of, I would tell you, resources, political situations, lack of personnel, et cetera, not everyone manages to comply with it. We have a substantial lapse of time in knowing if a vehicle is stolen or not." He elaborated,

> In theory, everyone should be analyzing their information on a daily basis. . . . But it doesn't happen. There are states that update on a daily basis. There are states that update every twenty-four, forty-eight hours. There are states that wait fifteen days up to a month. . . . Let me give you an example, from the State of Mexico, one of the largest states in the country. It's the state that has the largest number of stolen vehicles. In other measures, it also has a high level of crime. The State of Mexico receives the greatest number of federal resources, in this case, with the [Support] Funds for Public Security. They receive more money than the Federal District. And it is one of the states that generates some of the greatest amount of state funds by itself. And they have these types of delays. But I should also mention Tlaxcala. It has very little money, being a very small state. The population is small. So, the formula that exists for assigning resources leaves it receiving a tenth of what the state of Mexico receives. However, when the will to work is there, results can be achieved.

The lack of desire to achieve results was felt at the local level too. Matías Luna at REPUVE Zacatecas proffered that one modification he would make "would be to incentivize the personnel. I'm not talking about money. I'm saying there should more supervision of how you're doing your job. . . . And there should be supervision. It's a form of incentivizing the personnel to work in this program. . . . You've been here a week and you haven't seen anyone come to say, 'Hey, how are you doing?' I'm not saying they should be all over us here. But once or twice a week, have someone come to check. This incentivizes the personnel."

4.5.3 Political Interests

At times, the complications with implementing the REPUVE at the state level reflected not a lack of will on the part of state authorities but competing political interests, Daniela Flores's third point. As Samuel Gallo bluntly told me, "*Hay colores,*" or "there are colors," by which he meant political affiliations at work.[b] "It's like the case of Sinaloa," he explained.

> Sinaloa is drowning in crime and all that. But it's *priísta* [ruled by the PRI, the Institutional Revolutionary Party] and it doesn't share the ideas of *panismo* [of the PAN, the National Action Party]. The guy who became governor of the state was one of the principal detractors of the idea that the REPUVE had to have a chip. . . . Now the program is stalled because the governor was one of those who at the time didn't accept the passive chip. He wanted it to be active. He was going to sign a contract with a LoJack type of service, that type of satellite service, which has a higher operating cost, much much higher cost, than what we are implementing.

"Look," Gallo continued, "there are many questions that get in the way. I can tell you that Sinaloa is serviced now by the technology supplier who the Executive Secretary [SESNSP] is suing for having registered the website REPUVE.com. So there is a lawsuit launched by the Executive Secretary against this supplier, which is the supplier for Sinaloa. So there are interests. There is a conflict of interests." According to the State and Federal Operations directorate, the active-tag provider to whom they were referring—a Mexico City company named

---

b. In Mexico, as in many parts of the world where literacy rates have historically been low, political parties are often identified by colors or objects as much as by name. The Institutional Revolutionary Party, as the hegemonic political party following the Revolution, was able to wrap itself in the colors of the national flag: red, green, and white. The right-leaning National Action Party, which held the presidency from 2000 to 2012, adopts blue as its primary color. The third major, although comparatively smaller, party, the Democratic Revolution Party, colors itself yellow.

Socom—had publicly registered the name REPUVE in 2007, which was after the signing of the REPUVE law but before the decision was made on chip design and contractor.[48] In essence, the disputes over passive and active chips and the awarding of the REPUVE contract were continuing to complicate the implementation of the program. This situation underscores the variety of interests at play when "colors" are involved.

REPUVE administrators saw such conflicts of interest as endemic to Mexico. When I asked about the adequacy of resources for states, Gallo responded, "There's money. But it's poorly distributed. And the problem of distribution comes down to political networks. So, this is the problem. In those networks, someone is the mayor. Below him are his buddy, his son, his godson, and all of the family is occupying the political system." Daniela Flores added, "They are the ones who take the salaries. And that's where all of the money stays. They get $60,000, but it's divided between the same family." "And that's how you create these empires," Gallo continued, "this is how the monopolies are created. And when they leave power, new businesses and consortiums emerge. So this is the problem in the country. Because there exists this circle of impunity. Every three or six years," the length of time that someone serves in office in Mexico, "there are new rich people. This is impunity. There are no accusations. There is not legitimacy. Because if I accuse you today, I get kidnapped tomorrow. And they're going to split my head open. So why would I make a complaint?" Gallo and Flores reference here the personalistic political networks that have defined Mexican political life since at least the establishment of *caciquismo* (rule by local political chiefs) during the conquest.[49] While its forms have surely changed,[50] clientelism remains endemic in Mexico,[51] and democratization has thus far proved incapable of rooting out such "old corruption."[52] These political formations complicate the federal government's efforts to change the governance of automobility in the present day.

4.5.4 *The Organization of State Power*

Even when the resources and will to advance a program are present and deleterious political interests are absent, other political forces can intervene. For instance, the obstinacy of state governments to carry out the plans of federal administrators is provided for under the law. Within Mexico's federalist political system, which reflects the country's long history of dividing political power among regional strongmen,[53] federal law does not apply to state governments the way that it does to private corporations or individuals. As Vicente Bautista with the Procedures and Citizenry Directorate explained, "The problem that we are having these days is that the private industry, the producers, 100 percent are participating. But the law requires them to . . . the states are autonomous. Free and sovereign. We cannot sanction them. We have to convince them. That's the challenge for us." When I asked whether the government also had the power to withhold federal funding, such as the FASP, to encourage states' participation, Bautista was quick to reply, "No, no. We have security funds from the federal government that are allotted to them [the states]. But they decide how to distribute them according to their own priorities."

"This is one of the gaps," Samuel Gallo conceded. "It's something the law didn't take into consideration. The law stipulates that the factories, the importers, the law can grab them by the throat. It has them in its grasp. It can fine them if they don't comply. That part is taken care of. But since the constitution establishes that the states are free and sovereign, it isn't easy to establish these types of sanctions. It would never get passed by the Congress."

The Citizen Identity Card faced a similar obstacle in the organization of political power in Mexico. The CEDI ran up against existing state bureaucracy—that of the IFE and the voter card—established in an earlier era to fulfill similar functions. The duplication in function, and the threat posed by the new program, fostered political entrenchment aimed at preserving the IFE's authority. As IFE president Leonardo Valdez Zurita explained in opposing the CEDI, "the Card will negatively affect two of the pillars of our political system: the voter roll and the voter card" and even if the CEDI is "undoubtedly legal, from my point of view it's electorally unfortunate."[54]

In addition to the federated structure of power in Mexico, the Constitution only allows elected officials to serve single terms in office, a provision motivated by Porfirio Díaz's long tenure in power (1876–1911). For REPUVE program administrators, this constitutional guarantee against dictatorial rule created political turnover at the state level that affected the program's success. In Zacatecas, for example, which had been one of the leading states in applying RFID tags in 2010, a change from a Democratic Revolution Party (PRD) governor to a PRI governor was felt to disrupt the state's progress with the program. As Ignacio Meza from REPUVE Zacatecas delicately put it to me,

> I don't know if you have read about it or seen it, there was a PRD governor here in Zacatecas that changed to the PRI, the Institutional Revolutionary

> Party. There's a certain rivalry there. We don't see each other as citizens, but as enemies. And this has put the brakes on the process a bit. Other factors, such as the economic situation, have had an impact as well. This is a program that has to carry two enormous weights on its back. One, the federal government is *panista* [controlled by the PAN]. The other, it [the program] was begun by a *perredista* [PRD] government. So this hasn't permitted the push that these programs need, in contrast to San Luis [Potosí] or Veracruz. In Veracruz, the current government is the one that launched the program. There hasn't been a change of government. In San Luis, the same.

For the program's national directors, such cases spoke to larger challenges in the country. "The sad history," explained Samuel Gallo, invoking the legacy of *presidencialismo,* or the PRI's concentration of political power in the executive branch,[55] "is that whenever someone gets to that chair [presidency, governorship], everything changes. The problem is the chair. There is the sensation of power. Then all the good promises, all of the hopes that were built up, they begin to dissolve. They are not able to put into action even 50 percent of the campaign promises that they have. And now, in the next elections, there will be another change."

Instead of continuing a previous administration's policies, new administrations, Lucas Espinoza and Daniela Flores felt, resorted to populist tendencies and pursued more visible public works to win the support of voters. "Something unfortunate about the government here is that they want to show results quickly," Flores told me. "But you have to start from zero and it's going to take years, ten, fifteen years maybe [to get something done]. And they're not interested in that. Most of the alternatives that an administration has available to it are to take the populist route. It's better to construct a bridge than a highway, which looks sensational. So I'll get you a park. I'll get you a bridge. And that's that." "But then there are people who'll tell you," Espinoza half-jokingly interjected, "we don't have a river [for the bridge to transverse]. "Ah, well," Flores took back the thread of the conversation, "let's make a river too then! These are the types of illogical situations that there are." Placing themselves within the country's historical-political context, the REPUVE's directors saw their program as a victim of populist tendencies produced by political turnover.

As Flores's comments suggest, this political dynamic was present not only at the state or local level but at the national level as well. At the REPUVE national offices in 2011, a year before the end of Felipe Calderón's six-year term and a presidential election that most thought would be won by the PRI,[c] a mix of frustration and concern for what a change in "colors" at the top of the political ladder might mean for the program was palpable. "Unfortunately for us," Gallo shared, "we are a year and a half away from a change in federal administration. So important decisions are not going to be made. No one is going to dare tackle a problem or confront an institution because it's not going to matter. So, now, all the activity is going to start to go down and that is going to continue until the middle of 2013 and then begin to rise again. This is the change of administration. With a change of administration, new brains arrive. New ideas. And the continuity is broken."

The preceding points illustrate how forces beyond the actions of human actors, be they individual laypeople, corporations, or political parties, have affected the destiny of security surveillance programs in Mexico. The absence of adequate funds with which to launch programs, budgeting procedures for the expenditure of funds that do exist, nepotistic political practices that date back to colonial and pre-Columbian times, the arrangement of federal and state political authority and limitations on tenures in political office enshrined in Mexico's Constitution, and the presence of extant state authorities and programs created during earlier periods all illustrate the manner in which the structure of the state—the existing arrangement of rules, cultural practices, and governing bodies for conducting public affairs—can impede the establishment of new modes of governance.

4.6 TECHNICAL IMPEDIMENTS: DESIGN, SCOPE, AND GLITCHES

The challenges that the National Registry of Mobile Telephone Users, Citizen Identity Card, and Public Registry of Vehicles have encountered demonstrate that obstacles in political and state affairs emerge from multiple sources. This point is even clearer in light of the multiple technical difficulties the programs have faced. Three types of issues have loomed especially large: program design, project scale, and technical malfunctions.

c. And indeed, in July 2012, Enrique Peña Nieto, former governor of the state of Mexico and member of the PRI, was elected president by an electorate tired by the previous administration's focus on security issues at the cost of broader social and economic concerns.

4.6.1 Program Design

Poor program design hampered the RENAUT in particular. Despite popular resistance to the program, some eighty-two million mobile numbers were ultimately registered with the government,[56] representing 90 percent of all mobile lines in the country. Amid reports of dubious registrations, however, the government turned to verifying or inspecting the associations between phones lines and individuals. This would, in theory, be accomplished by comparing the names and dates of birth or CURPs associated with a particular number to the National Population Registry (RENAPO) maintained by the Interior Secretariat. How that task was to be accomplished however, or who would be responsible for it, was not clear.

Government officials, such as Mony de Swaan, president of the Federal Commission of Telecommunications (COFETEL), claimed that mobile telephone service providers were supposed to complete the work. "The registry itself has been successful. There are substantial numbers [of registrations]," de Swaan explained. "But where we have not been successful is in linking these registrations with the biometric characteristics of the user. We are in a stage where the registrations have to be linked with fingerprints. We need an analysis to know if this is working or not, because COFETEL cannot do the analysis. It's not our job. We don't have the capacity to do it."[57]

Industry representatives, on the other hand, countered that they had no knowledge of such requirements. Jorge Arreola, compliance director of Telefónica México, told the press that "the problem we have is with respect to the mechanism by which a CURP can be used to register a phone of some other person," the problem of the ten thousand Calderóns noted at the beginning of the chapter. The verification of identity was supposed to be undertaken by the Interior Secretariat, Arreola claimed, but "this project didn't move forward and we don't know if there exists a procedure that would allow us to move forward."[58] The uncertainty and lack of a clear plan to carry out the second stage of the RENAUT threatened the feasibility of the program moving forward.

Program design also affected the REPUVE. In 2001, before the REPUVE was born, the SESNSP, the governmental body that oversees the registry, moved from the Interior Secretariat (SEGOB) to the Secretariat of Public Security (SSP), a new agency created by then president Vicente Fox to fight crime. At the time, the SESNSP had responsibility for various technical operations, with direct control over the creation of databases related to security. During the Calderón administration, the SSP became the central pivot for security operations, including control over Plataforma México, the information system hosting and integrating various government databases; and the SESNSP returned to the SEGOB in 2009, on the logic that vital security functions should be housed within the Interior Secretariat. In the process, members of the REPUVE project felt like they lost contact with the technical support in the SSP needed to complete their work.

"In 2010, we as REPUVE ceased to belong to the Secretariat of Public Security and we went to the Interior Secretariat," Lucas Espinoza told me. "Over there stayed the technical part. So we lost another part of our history. And now, as the Executive Secretariat [of the National System for Public Security], we don't have a technical area that serves us, we don't have a technical area for development that can attend to the specific needs of this institution. All the time we have to ask favors from Plataforma México." Daniela Flores added, "And we have to stop everything for them to notice us. We moved from being a complete administrative unit to being third-party clients."

The loss of access to technical solutions, as Flores hinted at, resulted in delays. "We have had a series of complications with the technical aspects," she explained, "They [Plataforma México] don't have the capacity. . . . So I enter a queue, and they are assisting everyone else. I have a spot in the line and however hard I try, they can't get me the requirements like I need them." The delays, in turn, complicated the State and Federal Operations team's interactions with the states. Flores continued, "So we can't give the states adequate tools, so that everything keeps working. And then they lose interest." "We have the [software] applications now," Samuel Gallo interjected, "but they were two years late. If we had had these applications last year, we would have implemented more than 50 percent of this project. If Plataforma México had given me the applications when I needed [them] last year, we would have 50 percent of the states working [with us] and I wouldn't be worried about anything more than six or seven other states. In 2012, we'll certainly lock them up. But these are the gaps."

These concerns over the organization of work and program design appeared at the local level too. In Zacatecas, technicians inspecting vehicles complained that they did not have access to the stolen vehicles database to facilitate their work. Instead, they had to call over to the prosecutor's office to access the information. Matías Luna grumbled that the software "doesn't have some of the features that would help the work go easier. For example, we have to do a report, a report of not being stolen when we are verifying the VIN [vehicle identification number]. We have to pull a report, but we have to depend on another office in order to be able to pull it. Why can't we use the same software so that, when we type in the VIN, it can tell me that or give me the report automatically?"

These comments from those on the frontline of REPUVE's rollout in Zacatecas and those working in the SESNSP's national offices suggest limitations in *prohesion* as a mode of governance. As governmental agencies are increasingly integrated through centralized electronic databases, access to those centralized information networks becomes indispensable for those agencies to do their work, while restricting access is fundamental to maintaining security. A rivalry develops, between the desire for the state's bureaucratic tasks to be completed in a timely manner and the demand for the information the state is administering to be safeguarded.

4.6.2 Project Scale

Even when the programs were operating without major complications, they were challenged by the immense scope of the work to be completed. The State of Tlaxcala, for instance, held out earlier by Samuel Gallo as a model for the states' implementation of the REPUVE, was able to apply chips to some twenty thousand vehicles within six months of beginning operations at two REPUVE sites, an impressive rate of forty vehicles per day. However, that success represented only 10 percent of the state's vehicular roll, leaving the program years short of covering the entire population.[59] To provide another example, the State and Federal Operations directorate described how it had succeeded in creating a centralized database from the vehicular data from all thirty-two federated states. But the enormity of the task cost the office years to complete. Gallo explained,

> We asked the thirty-two states, "Give us your databases," and we reviewed them, one by one, taking out trash, taking out duplications, all of this, in order to have a trustworthy database. We were 100 percent dedicated to reviewing each of the registries so that they were complete, so that they weren't duplicated, so that they didn't contain trash. [But] one problem was the great diversity in record keeping. We would find cars that were registered in three, four, five different states. They hadn't been removed [from other

states' vehicle rolls]. Sometimes, they had typed in Volkswagen and put *b* in place of *v*, and they would write Ford without the *d*.[d] So we had a lot of data-entry errors. Lots of trash. Omissions. This process took us three years. We took three years to clean the database.

This editing is precisely the type of cohesive work expected of the REPUVE in integrating the data systems of diverse entities. But the sheer scale of the task, combined with the limitations of Plataforma México and a lack of resources, left the team behind with other work.

The scale of the REPUVE's work involved geography as well. In Zacatecas, a state with an appreciable rural population spread over a large geographic area, ensuring that vehicles in rural areas were included in the program posed a particular challenge. "It's like with the license plates," Diego Avila told me. "We know that in Zacatecas there are a half million registered vehicles, but there aren't a half million vehicles. There's more. Some aren't plated. Others have American plates. But there are more than a half million vehicles, which means that not all of the vehicles in the state are taxed." When I inquired why or how people don't register their cars, Avila responded,

> Where are these people? In the villages. Why? Because maybe they make only a trip or two to the city. In the towns, in the cities, it's a bit stricter, because there's more people watching. There are cameras. You can't get away with as much. Eventually you're going to come across a traffic cop who's not going to do you a favor. And they're going to stop you and bring your car to the tow yard because you don't have plates. But there are villages that are very small, that have one or two traffic cops at most. And then it's your buddy, your neighbor, your brother, or your father who's stopping you. And then it's just "go on your way" [and nothing happens]. There are a lot of vehicles like that.

Interestingly, the same close-knit, local political networks that Daniela Flores and Samuel Gallo complained prevented them from implementing the REPUVE from above are also seen to work against the program on the ground.

Describing the geographic challenge for Zacatecas, Ignacio Meza, the state official responsible for the REPUVE there, explained that "the problem with Zacatecas is that we have towns almost four hours from

d. In spoken Spanish, *b*'s and *v*'s are interchangeable and hard consonants at the end of Hispanicized foreign words are often dropped. This example is a reminder of how the unruly nature of language can complicate controlling human communication.

here [the capital city]. For example, to the north is Concepción del Oro and Mazapil. They're small, but we have to provide them the service." So the state decided to employ mobile registration modules, housed in mobile trailers that "could be moved for a week or two, to have a fixed module there." But in electing mobile units over stationary ones, the necessary computer software had to be broadcast remotely via radio frequency. "At the time," said Meza, "I spoke with the data specialists, [explaining] that this was the model we were going to replicate in all of the principal towns. We were not going to be able to connect all of them with cables. The idea was to install these mobile modules. . . . So we tried to replicate in Zacatecas how the connection would be there, via remote antennas." But frequently during my observations, the team registering vehicles experienced service disruptions that delayed the transmission of data from the registration site to the secretariat's database. When I asked Meza about these disruptions, he shared that the system specialists "said that these things are out of our hands. Maybe the telephone company, which administers the Internet connection, had an issue. Or maybe there was a problem of some sort. Someone hit a telephone pole. Or someone was digging and cut the fiber-optic cables."

4.6.3 Technical Glitches

The difficulties experienced by REPUVE Zacatecas in providing ambulatory service touches on another challenge facing the Mexican government's attempts to enroll surveillance technologies in the fight against crime: technological malfunction. Prohesion as a form of governance depends on the performative capacities of surveillance and other information technologies. When those technological artifacts cannot do what they are intended to do, the programs themselves are at risk.

Technological malfunction can be caused by a struck telephone pole or an unknowing utility worker who cuts a ground cable. Or it can result from other dangers lurking in both built and natural environments. Sucaro, for instance, the car importer, had to figure out how to tag vehicles coming off ships from Asia en route to their dealerships. "Distinct from other companies that have their own yards," Agustín Sandoval and Felipe Ibarra explained to me, "we have a process where on the ship that arrives to port, some of the cars go to a yard as inventory, but the majority have already been sold. So they go directly to the dealers. The challenge was to set up the application of the chips without affecting the commercial process. The solution that we had to come up with was to do everything at the port." But the company soon encountered problems, according to Sandoval:

> The first big problem we had there was that the printer bar codes from the plant arrived pixelated. The print was a bit blurry. And when there is a lot of sun, which is normal, the infrared couldn't read [the bar codes]. We struggled with this. So we would begin the work very early, at eight in the morning, and it [the tag] could be read because there wasn't so much sun. But when the sun really started, we would start to have problems. To find out what was going on, we had to get in touch with the producer of the handheld readers directly. And they told us the handheld readers had to operate with thermal paper. Without it, and with extreme solar radiation, it wouldn't read.

If the technological artifacts on which the Mexican state's prohesive plans depended could fail at inopportune times, the things they were meant to monitor could present challenges of their own. For the REPUVE, for instance, variability in vehicle identification standards has complicated vehicle registrations. There is no single standard for assigning VINs. The seventeen-digit number identifying the car's manufacturer and characteristics, including model year, was adopted in many parts of the world in the early 1980s, though the United States and Canada employ a slightly different protocol. In Mexico, as Rodrigo Domínguez with REPUVE Sonora told me, "the international norm, or the norm from America, was applied . . . beginning in 1997. All of the manufacturers—or let me put that in quotation marks—most of the manufacturers have followed that. And this has been the norm really since 1997." Vehicles produced in Mexico before that time, however, present particular difficulties. "Nissans and Volkswagens from '95, '96, '97, as well as Chevys from '94," Domínguez elaborated, "they can't be put into the Public Registry of Vehicles for now. The system doesn't allow the registration of their serial numbers because of a production problem in those vehicles. The REPUVE system detected duplicate serial numbers in those brands, which caused the service to be suspended."

The differing norms for identifying cars complicates vehicle inspection as well. To verify a vehicle's identity, REPUVE inspectors must document three matching instances of the VIN. The first instance is routinely located on the vehicle dashboard. The second is commonly located on the car body, under the hood and below the windshield wipers. A third is usually found on the doorjamb. If the VIN is absent from any of these three locations for whatever reason, the next place to look is on the chassis. And locating that VIN can be problematic, as it is not intended to be easily found. Ignacio Meza with REPUVE Zacatecas told me, "It took us fifteen days to be trained with all the makes and models. In addition, they [the technicians] have a logbook where they note the make, model, and where the secret numbers are located." This logbook was a point of pride among the inspectors, who approached this aspect of vehicular verification as a sort of riddle or problem to be solved.

Of course, each day presents its own challenges for inspectors, and a vehicle not previously encountered can always appear.[e] "Just yesterday," Rodrigo Domínguez, the REPUVE Sonora site manager, told me,

> a Nissan Tsuru came by that was completely different from what was on the bill of sale. What the bill of sale said was that it was a Ford Explorer. What happened is that the car had been hit, they recovered it, and they reassembled it with other pieces. So how am I going to validate it? I said to the guy, "Give me your documents, your bills, et cetera," and with that we were able to document that we had encountered a vehicle with a different serial number. In this case, the technician recorded the serial number from the doorjamb to register the vehicle. . . . The vehicles can receive a chip as long as we normalize [these discrepancies]. But if nothing remains of a car other than the parts, or 50 percent of a vehicle remains, then it's the other 50 percent of the other vehicle that we would have to register.

If automobile idiosyncrasies can present challenges for their registration and inspection in the REPUVE program, so too can they complicate their regulation. "Another problem that came up, which was a big pain," explained Domínguez in Sonora, "was the fact that vehicles that have metallic particles [in their windshields] don't allow readings from the equipment. They estimate that around 3 percent of the 25 million vehicles in the country, around 750,000, cannot have readings. . . . One of these is the Partner from Renault from 2007. It is going to be a pain for owners, unfortunately. We cannot put a chip on the windshield because it won't be read." Metallized windshields in certain makes and models are common. And while REPUVE administrators hoped that the number of vehicles affected would be less than 3 percent of the total

---

e. During my week at the REPUVE site, a Jeep Liberty arrived for its sticker. The team had never encountered the vehicle before and was having difficulty locating a third VIN. Being the son of a Jeep dealer, I decided to spend a not insignificant sum of Telcel credit to dial the family dealership and ask where the third number might be found. The dealership had the location for us in a matter of minutes, suggesting perhaps that greater coordination with automobile manufacturers or dealers could mitigate the challenges of locating VINs in the future.

vehicle roll, in fact more than thirty makes and one hundred models have metallized windshields.[60] Thus, the reading of REPUVE's RFID tags has been complicated by the materiality it is meant to control.

In addition to vehicles, paperwork can also present complications. Generally, drivers need to present five documents to register: a bill of sale from the purchase of the vehicle, a vehicle registration, a driver's license, a proof of residency, and an official form of identification. At times, people I spoke with who were registering their vehicles had difficulty gathering the necessary documents to obtain their tags. "I had to fight a lot for the documents they ask for," one woman told me in Sonora. "I fought. It was annoying because they said that the documents were fine. But afterward, they told me that they were insufficient and I made three appointments to come in. So I had to come in three times because of misunderstandings about my documentation. The last time they told me that I was missing this document and that. I was sick of it. Finally, I said, 'Hey, I'm not going to get anything else.' That was the last fight and it looks like it's OK now."

Some drivers with paperwork problems blamed themselves. "It was my fault," said another woman. "I didn't bring the required documents. So I had to make three trips. [Three?] Yeah, three, but they told me. They were very clear with me at the beginning in saying, 'You need this, that, and the other thing, the original bill of sale.' I didn't want to bring it because it's a bill of sale and I keep it safe. Now I brought it and everything is perfect."

But others were not so understanding. "It's incompetence. How do you say it?" complained one man I spoke with at the module in Sonora. "Negligence":

They have to verify the documents. I came yesterday to complete the transaction and they told me I lacked a document. I brought it today, and today they tell me that I am missing a notary on my bill of sale. This is a pain. Because yesterday they should have told me, "You are missing a stamp. Come again tomorrow and have them stamp this." Now they tell me that they can't do the transaction for me because I have to bring in the notarized version. I asked them, "What is your job here?" She told me, "We don't stamp that." Fine, "But what is your job? Checking that you have all the documents and that the documents are in order. But you didn't do that." She started with, "Well, what happened is . . . " "No! You didn't do it. If you told me yesterday that the document was missing a stamp, I would go today, or go yesterday, and have them put the stamp on the document. And today I'd come here and you would finish the transaction." Now I have to go to another office so they can put a stamp that they didn't put before.

The REPUVE's diverse technical challenges discussed here, it must be noted, did not necessarily jeopardize the registration and inspection of vehicles. Sketchy network connections could eventually be reestablished. Someone low in the Plataforma México queue would eventually receive service. Erroneous data entries could be cleaned up. Rural areas outside the grasp of the law would eventually be reached. Malfunctioning equipment could always be diagnosed and fixed. Hidden VINs could, with persistence, be unearthed. And missing paperwork could eventually be supplied. But these challenges each led to delays that increased dissatisfaction with the program among drivers, *sujetos obligados,* and governmental actors who were choosing to stand "before the law" and participate.

4.7 RETHINKING RESISTANCE; OR, HOW COLLECTIVE AGENCIES GIVE THE PROHESIVE STATE THE SLIP

And so it was during 2010–11 that the Mexican government found its plans to gain a firmer grip on mobile telephony, automobility, and personal identification floundering on the brink of failure. Individuals fearful of losing their privacy or entrusting their data to the state organized themselves in digital spaces to share strategies for evading the telephone registry. Car companies complied with new requirements to facilitate the automobile registry but launched a lawsuit and took to the media in an attempt to modify the REPUVE legislation. The federal agency that saw its mission threatened by the new national identity card worked the media to foster opposition to the card. State governments apprehensive about the costs and potential for success of the vehicle registry, meanwhile, either refused to apply tags to vehicles or suspended operations. And those states that did apply the tags found themselves, like the *sujetos obligados* in the private sector, facing technical challenges that complicated their attempt to comply with the law.

These difficulties experienced by the Mexican government can partly be explained, as this chapter has shown, using the conceptual repertoire that political scientists and sociolegal scholars have developed to study the state and law. Ordinary Mexicans opposed to the RENAUT organized themselves to "stand against the law." Their actions did not attempt to repeal the law or topple the Calderón administration. Rather, they, like the *campesinos* who refused to participate in Mexico's Agrarian Census following the Revolution,[61] attempted to evade the law either by refusing to register or by providing false information, individual actions that can be read as "weapons of the weak." The decidedly nonweak—multinational automobile corporations—by contrast felt that they could not risk being so bold or flippant in their position vis-à-vis the law. Not complying with the REPUVE risked state sanction, which could harm business operations more than complying with the law in the first place. Nevertheless, the new automobile registry did imply significant costs. Thus, the companies opted to challenge the program "through the law" by suing to have the law changed.

But these conceptualizations of resistance only go so far in explaining the difficulties facing the Calderón administration. Major challenges jeopardizing these programs came not only from individual citizens and incorporated businesses. In the case of the Citizen Identity Card, the main opposition originated from within the state itself, the Federal Electoral Institute. In the case of the automobile registry, the primary sources of defiance again arose from within the political system, the individual states or *entidades federativas* that refused to implement the program. In the case of the telephone registry, in addition to the ten thousand Calderóns, design failure was the major obstacle that left the government without a clear strategy for verifying the phone numbers of the 90 percent of mobile phone users who did register their data, whether earnestly or not. Thus, the trials and travails of the Mexican government in implementing its prohesive vision of governance suggest sites of conflict and contestation beyond those that have captured the attention of scholars in the past.

With regard to the state-based obstacles encountered by the CEDI and REPUVE, it is interesting to consider, if we take a long enough timeline, that the Federal Electoral Institute's voter card, or even the federalist composition of political power in Mexico, embodies the substance of earlier efforts to manage the collective agencies of society. As noted in the first chapter, the IFE is a body created by Mexican authorities following the popular outrage over the 1988 presidential election, the one in which IBM's computers crashed as it appeared that the opposition candidate, Cuauhtémoc Cárdenas, was heading to victory. And the voter card and the biometrics it contains took shape over the twentieth century in order to govern personal identity in a manner conducive to democratic governance in Mexico. The federalist system, meanwhile, enshrined in Mexico's Constitution following the bloody civil conflict of the Revolution, is a solution to the challenges of constructing a single, unified political system for a geographic space defined by regional rivalries since pre-Columbian times.[62] In this sense, the challenges encountered by the Calderón administration in enacting its new strategies for capturing the collective agencies of Mexican society are concentrated in the old infrastructure, bodies, practices, and agreements that had been developed to capture the collective agencies of Mexico in the past. This is, in essence, "law against itself."

A similar dynamic can be detected in the technical challenges faced by the REPUVE and RENAUT programs. With cars manufactured before 1997, when Mexico adopted an international standard for identifying vehicles, the REPUVE comes up against an older way of ordering vehicles that does not lend itself to the new strategy. And these vehicles are able, then, to escape the grip of the new registry.

But with automobiles, a challenge beyond older modes of ordering is their materiality. Vehicles resist government efforts to be controlled. Metallized windshields provide a material barrier that insulates cars from being read. The absence of windshields or the reduced size of windshields on motorcycles, meanwhile, has prevented them from being tagged by the REPUVE program. Reassembled cars, a common phenomenon in Mexico and other poorer countries,[63] a technological embodiment of *mestizaje* (racial mixing), befuddle official ways of identifying vehicles to guarantee "legal certainty."

Of course, the material objects complicating REPUVE's operation have not just been automobiles. Computers crash, their operability stretched by the need to communicate remotely in the rural expanses of Mexico. Bar codes pixelate when printed on nonthermal paper, hiding the information they were meant to convey when the sun intensifies. And other technological challenges—disruptions in mobile telephony for Telcel subscribers or unresponsive webpages—are an ever-present reality in the material world.

These examples invite a reconceptualization of resistance that takes account of such forces. But how would one do that? A sensible place to start would be science and technology studies (STS) that highlight how nonhuman things—sea scallops in Saint-Brieuc Bay,[64] the anthrax ravaging cattle in the French countryside during the nineteenth century,[65] the bubble chamber built by Donald Glaser to detect subatomic particles[66]—"resist" the efforts of scientists and engineers to bring them into their plans for ordering the world. This "material agency" is at the heart of scientific and technological failure, just as scientists' and engineers' ability to "capture" it lies at the heart of its progress.[67]

These concepts can be brought to bear on surveillance and the state. Like scallops, anthrax, and bubble chambers, the material agency of windshields, bar codes, and computers represent points of contestation that challenge the power and operation of state surveillance programs in Mexico. But we need not stop with material objects. Rather, the nonhuman agency working against the CEDI and REPUVE includes certain arrangements of political life or ways of ordering society—the "co-productions"[68] or "assemblages"[69] of the state—that are inherited from authorities' past attempts to capture the collective agencies of society.

To formalize this train of thought, we need not limit ourselves to defining resistance, as James Scott does, as "any act(s) by member(s) of a subordinate class that is or are intended either to mitigate or deny claims made on that class by superordinate classes or to advance its own claims vis-à-vis those subordinate classes."[70] This definition restricts itself to human actors of a subordinate class, overlooking other force relations in society that can count more centrally in the fate and outcomes of governmental projects and define the contours and limits of power. Instead, we might better define resistance as "any force, whether human or not, that has the effect of obstructing the intended plans and intentions or established relational patterns of authorities."

This definitional change provides both a fuller understanding of the experiences of surveillance technologies in Mexico's War on Crime and a different way of studying surveillance and state power going forward. For one, it shifts analysis from "members of a subordinate class" to "the plans and established patterns of authority." That is, it moves analytical focus from the weak to the strong. The rationale for privileging the actions of the poor is clear for a field of research interested in contesting social inequalities. "The celebration of some forms of resistance contains implicit commitments to social justice and equality," Sally Engle Merry writes. Thus, "it would be more honest to acknowledge where we stand and join in the search for a more just world."[71] Practically, however, we can ask whether focusing on the weak leaves the powerful out of our immediate focus. To understand power, it seems right to place the designs of the powerful at the forefront of analysis.

Second, the new definition shifts analysis of resistance from relativistic meanings (the "intentions" of individual members of subordinate classes) to general outcomes ("the effect of obstructing plans" that affect the social environment through which human activity is conducted). The reasons for wanting to privilege the intentions of ordinary people are again understandable. The disappointments of collective action following the 1960s and 1970s in the United States, Europe, and Latin America owed in good measure to social movements' disregard for the views, values, and interests of some members, usually those who were not male or European, which limited the movements' appeal and democratic potential.[72] Too often, the means were ignored for the ends. Nevertheless, attending to the intentions of actors also draws our gaze away from the more objective, material outcomes of action that, as STS has shown, must be taken into account to understand social action and outcomes.

If the preceding two points, by moving away from perspectives that prioritize the meaning making of the less privileged, seem conservative, it bears saying that reconceptualizing resistance in this way can also expand what is considered the field of political dispute. By limiting ourselves to a definition of resistance as "intentional action,"[f] our analysis can overlook those actions that might be unintentional but still effective in countering the plans of authorities. I have in mind here some of the tweets against Mexico's phone registry that started this chapter. A tweet such as "They can take away our cell phone numbers but they cant [sic] take away our freeeeeeeeeedooooooooooom" or "Did you know the #RENAUT plans to include your fingerprints in a second stage?" can only with difficulty be read as intentional actions in opposition to the government. The first is a stab at humor, a play on Mel Gibson's iconic rallying cry as Braveheart, and the second could merely be informational, telling readers what will follow with the registry. However, these comments, something akin to gossip or chatting, can diminish the legitimacy of the registry and serve to counter its power. Similarly, simply purchasing a vehicle that turns out to have a metallized windshield or reassembled parts can unknowingly obstruct the advance of a governmental program like the vehicle registry.

These actions cannot be described as democratically expansive, but they do encourage us to imagine a wider space of political contestation than might otherwise be imagined. Indeed, if collective agency in society can be seen as distributed across a collection of human and nonhuman actors, including governmental state structures that are "co-produced" in the process of regulating those agencies, it stands to reason that each link in that distribution can serve as a site of resistance. In Mexico's War on Crime, as the federal government has sought to reorder mobile telephony, automobility, and personal identification by "making things stick," these agencies have proven resistant to change at points all along their distributions, from the intentions of ordinary Mexicans to the pixels of bar codes.

f. Jocelyn Hollander and Rachel Einwohner note that the common threads between the disparate uses of "resistance" in the social sciences are "a sense of action" and "a sense of opposition" (Hollander and Einwohner, "Conceptualizing Resistance," 538).

CHAPTER 5

Statecraft

The chip is not mandatory. There is no law that requires
people to get a chip. The motivation [for people to get the
chip] is passing through the toll [area] free of charge with
the tag.

—Carlos Aguilar, REPUVE Sonora employee

The chip helps me avoid wasting time in line like all the other
vehicles. You arrive in your lane, and the reader automatically
reads the chip and lifts the gate. So it saves time.

—REPUVE registrant in Sonora

5.1 STATES OF SUCCESS

During December 2013, an odd daily ritual took shape at the Public Registry of Vehicles (REPUVE) installation at the old Santa Rita market in
Ciudad Juárez, Chihuahua. Each morning, dozens of local residents lined
up in hopes of securing a REPUVE chip before the end of the year. As
the days passed, and demand for the chip increased, the lines grew longer, and arriving early in the morning to queue for the radio-frequency
identification (RFID) tag no longer sufficed. Residents began camping for
nights at a time in freezing temperatures and without toilet facilities for
a chance to join the program. Describing her predicament, Martha Oliveros Hernández, a young mother in a green Dodge Caravan, explained
to reporters, "I arrived at three in the morning. I'm going to stay through
the night until tomorrow morning. I don't want to do it. But I need it [the
chip]. They say that it's important because next year it [the chip] will be
required and they are going to give fines. The cold is intense. But we're
going to spend the night all the same. We don't have an alternative."[1]

To meet the demand and leave parking spaces for drivers arriving to renew licenses and complete other transactions at the complex, REPUVE employees implemented a number system to allow drivers to hold their place in line. Each day, 120 numbers were distributed. But this failed to quell the queues, employees reported. "The same people come and line up despite [our] informing them that they don't need to. They have the number corresponding to them placed on their vehicle. We are taking measures so that only half of the parking lot is occupied and the other part is available for those people who are getting plates for the first time."[2]

The cause of this sudden demand for the REPUVE program was simple enough. As Ms. Oliveros alluded to, a rumor had circulated in Juárez that the chip would become obligatory in 2014, that it would have a cost, and that those without it would be subject to fines. "I came now because it's free," explained Guadalupe, another young woman spending the evening in her vehicle. "In the coming year, who knows?"[3]

The rumor was fed by employees at the REPUVE site, who had noted earlier in the year that sanctions would be imposed on those without chips. But now faced with the sudden surge of local residents desiring a chip, administrators sought to clarify that there was no need to line up for tags, since there were no plans to make the chip obligatory, more registration centers would be established in the coming year, and a publicity campaign was forthcoming to explain the program to drivers. "It is important to emphasize to the community of Chihuahua that there is not any rush to enlist in the public registry of vehicles," explained Idalí Mariñelarena, an administrator with REPUVE Chihuahua, "because drivers can come throughout the year without any problem."[4]

Given the various setbacks endured by the telephone registry, national identity card, and automobile registry from 2010 to 2012, the scene at the Santa Rita market might come as a surprise. But Juárez's experiences were not without precedent. A similar scenario had played out a year earlier in the state of San Luis Potosí, one of the early adopters of the Public Registry of Vehicles program.[5] The state of Quintana Roo, home to the tourist resorts of Cancún, had to expand its hours of service to the weekends to facilitate demand once it began distributing chips.[6] And the governor of the state of Coahuila faced popular pressure from civic organizations for failing to enlist the state in the REPUVE, a neglectful oversight they felt placed their property at risk.[7]

These vignettes help underscore that the Public Registry of Vehicles, while facing considerable challenges, has in fact met with some success. By summer 2012, nearly every automobile manufacturer and importer in Mexico was applying tags to their vehicles, which meant that 1,737,573 tags had been applied to new vehicles and 7,366,856 vehicles had been added to the national registry's database. The *sujetos obligados* (obligated subjects) contributed a mountain of digital files to the REPUVE database as well—financial institutions contributed some five million sales records; insurers, some twenty-six million notices of policy changes. The number of states applying tags rose from six in 2011 to fourteen in 2012.

But more significant was that the REPUVE had modestly begun to penetrate the daily lives of Mexican drivers. The call center established to handle people's inquiries about the status of their vehicles received 24,000 calls in the first half of 2012. A small cottage industry sprang up at *tianguis,* the open-air car markets with a reputation for dealing in goods of questionable provenance, where enterprising individuals with laptops and printers charged prospective buyers thirty pesos (about three dollars) to check the REPUVE website for the legal status of vehicles up for sale. State and local police also set up terminals outside the *tianguis* as both a service to car buyers and a deterrent to retailers of stolen vehicles.[8] And throughout the country, the REPUVE database became a centralized source of information about vehicles that police officers on patrol could use to identify stolen cars.[9] Thus, if the program was not fully operational across the country, there was a real sense that the REPUVE was finding its legs.

Something similar could be said of the Citizen Identity Card (CEDI) by summer 2012. In 2011 and 2012, some three million personal identity cards were distributed. However, in a modification to the program's original design, all of those cards were distributed to minors, such as Leslie García Rodríguez, the young girl mentioned at the start of the second chapter. In January 2011, the federal government, under the leadership of Interior Secretary Francisco Blake Mora, decided to issue personal identity cards to Mexicans under the age of eighteen and to postpone issuing them to adults, to accommodate the concerns of the Federal Electoral Institute (IFE).[10] To relaunch the card, which was rebranded the Identity Card for Minors (Cédula de Identidad para Menores), President Felipe Calderón appeared at a public event in his home state of Michoacán and championed the card as an achievement of the Mexican government in fulfilling the obligations of the 1990 Population Law.[11] As part of the relaunch, the card was recast as a security measure that would protect children from kidnappings and human trafficking as well as facilitate registering for school, accessing health care, or obtaining a passport. The Identity Card for Minors began being distributed at schools in the states of Baja California, Colima, Chiapas, Guanajuato, Jalisco, and Nuevo León. At a school in Baja California, María Isabel Miranda de Wallace, head of the organization Stop the Kidnapping, kicked off a ceremony titled "There is No One Like You" by telling the audience,

> The Identity Card gives us an official identity so that we know who we are before the state of Mexico and whichever authority. . . . My name is María Isabel Miranda Torres [maiden name]. I have the certainty that although another person may exist with the same name, that person doesn't have my fingerprints, doesn't have my eyes, and this gives me a unique identity. . . . And what I want to say to the parents of the families is that no one wants to risk having a child get lost and not be able to locate him. May we never, ever lose any child. May no child ever be subject to trafficking. The people who entrap on the Internet, sometimes they capture children by making them believe that they love them. But really they want to abduct them. May we never have to experience that. But it is important that we have this identity [card] because it would help locating them. If a child is lost, this will be the difference. Dear parents: This identity card truly has infinitely more advantages than whatever the political debate says. Don't let yourselves be fooled. Sometimes politicians . . . bring things into the political arena that shouldn't be brought in. This is a matter of security. This is a matter of security and safety of our children. May we come to implement this identity nationally, for the children and the adults.[12]

In contrast to the REPUVE and the Identity Card for Minors, by summer 2012 the National Registry of Mobile Telephone Users (RENAUT) was no more. In April 2011, in response to the lack of progress in verifying the identity of phone users who had registered with the program, the Senate voted to abolish the RENAUT.[13] While the program passed into history, organizations such as the Federal Institute for Information Access and InfoDF pressed the government to ensure the destruction of the RENAUT database, to prevent it from falling into criminal hands.[14] There also remained the question of what to do about the tracking of cell phones going forward. On April 19, 2013, the Interior Secretariat reported that it had completed destruction of the RENAUT database, thus bringing to a close the peculiar history of Mexico's mobile telephone registry.[15] But the question of what would replace the RENAUT remained unanswered.

These summaries of the REPUVE, CEDI, and RENAUT programs illustrate a fundamental point concerning state formation and the implementation of surveillance technologies to fight crime in Mexico.

If the technologies met with various types of resistance that threatened the designs of those in power, that resistance did not decide the programs' fates. Some programs and technologies survived and were slowly adopted across society, while others withered away. What explains these differences?

The simplest answer to this question is that many Mexicans simply saw benefits in some of the programs and not in others. In the REPUVE, a more secure and remunerating hold over vehicles could be imagined. In the RENAUT, by and large, a similar secure, promising future could not. But, this chapter argues, such outcomes are not given in advance and involve more than personal predilections. Central as well has been the manner in which program administrators have responded to the resistance posed by citizens, companies, and state officials. That is, the trajectories of these monitoring programs were a function not simply of how people perceived and reacted to them but also of how authorities reacted to those reactions. This chapter provides an inventory of the various strategies taken by administrators to overcome resistance to their programs. These include building *technical resilience* into the operation of programs to make them less onerous or more user-friendly for those obligated to participate; *providing quality service* to *sujetos obligados* and other individuals to address their concerns and illustrate program benefits; making *threats* and offering *incentives* to motivate peoples' participation; building a *critical mass* of participation to preserve and bulwark the programs against future setbacks; *allowing the levying of taxes* by state actors to incentivize their participation; *piggybacking* programs onto existing state infrastructures and institutions; and, when all else fails, *circumventing the state* altogether to build a new infrastructure of governance in collaboration with private actors.

Two lessons can be drawn from reflecting on these strategies for countering resistance to surveillance. First, the success of surveillance programs owes not only to the plans of authorities or the design of technologies. Rather, in good measure it owes to the dexterity of bureaucrats to go off-script and find strategies central to successful implementation. These bureaucrats engage in what can be called *statecraft,* practices in the art of governance that help give shape to the state.

Second, if improvisational tactics allow the Mexican state to get a hold on collective agencies, it is not the prohesive hold authorities had envisioned. Rather than using the RFID chip to instantaneously locate vehicles throughout the country, authorities in Sonora employ it as a tolling solution. In place of distributing identity cards to all Mexicans to simplify identification, the state only issues the CEDI to schoolchildren. And in the face of an ineffectual state bureaucracy unable to manage the RENAUT, the federal government turns to the private sector to deputize phone companies to maintain surveillance over mobile communications. Thus, the shape of state reformation through surveillance technologies cannot be known in advance, whether through the dictates of state leaders who adopt them or the investigations of scholars who study them. That shape instead emerges in time through the interplay of the resistance that arises to the technologies and the accommodations of *statecraft*.[a] In this sense, the state—despite its structured arrangement of institutions and roles, laws and regulations, authority over a fixed geographic area, and coercive technologies—is not a static entity, but a stochastic one whose interactions of power are patterned on the past but cannot be predicted or determined a priori. If this uncertainty concerning the impact of surveillance technologies on contemporary governance in Mexico leaves our social-scientific desire for certainty unfulfilled, it can also be read optimistically as evidence of the continued influence of human agency on the emerging technoscientific security state.

5.2 PERCEIVED BENEFITS

As noted above, the different outcomes experienced by the REPUVE and RENAUT owe to the fact that people liked the REPUVE and disliked the RENAUT. Returning to the survey I conducted, only 43.3 percent of those asked (42 of 97) supported the RENAUT, while 76.3 percent (74 of 97) supported the REPUVE. Respondents' answers help explain this difference. While the RENAUT, as noted in the last chapter, aroused fears about personal privacy and data security, the REPUVE connected to people's desire to protect their property. For example, respondents noted the following about the REPUVE:

I believe my auto is more secure from robbery and it would be
   easier to locate it.
You would know if the car was stolen or if they used it improperly.
This would help to more quickly locate vehicles that are stolen.

a. In explaining state formation through a "dialectic of resistance and accommodation," this chapter borrows from Andrew Pickering's description of science and technology as a "mangle of practice" (Pickering, *Mangle of Practice*).

This concern with protecting property aligned with what registrants in Zacatecas told me about their motivations for participating in the program:

> When they register the car, they check all of the documents, all the papers, to make sure they're in order. This ensures that all the cars have valid documentation, the documents for the car are in order. I think if they applied this to everyone, there would be more coordination over all the cars. There wouldn't be stolen cars.

> Simply because I go on excursions sometimes with my family to Aguascalientes, to get to Rancho Nuevo Morenos. That's where I go, I take my family in the truck. That's the reason. The insecurity that exists. [Criminals] will take your things whenever they want. Maybe with these chips that they put on, one can travel with more security. Maybe they'll be able to locate the vehicle.

> I'm here because of robberies. There is a lot of insecurity and they take our trucks. There is a lot of organized crime here. We live up in the villages, I'm a farmer. They come and hold us up with pistols and everything. The hope that we have with the chip is that they can locate [stolen vehicles] immediately.

> "I believe in the PAN [National Action Party]. The PRI [Institutional Revolutionary Party] scares me. And we're in for many years of PRI. In December [when the PRI would assume control of the presidency], the country is fucked. . . . This technology is state of the art, never before did we have chips or radar. Before, it was by the grace of God, pure good will. In addition to that, a month ago a lady hit me. I got her license plate, she left, but she left her plate number. Ten years ago, even with the plate number, it would have taken fifteen days to find the car. Now, in five minutes I had her name, and address. So, we're talking about this [vehicle registry] making things more agile.

These comments by drivers in Mexico express a clear confidence that the Public Registry of Vehicles, through its registration and inspection of vehicles and subsequent regulation through advanced surveillance technologies, will help ensure the security of one of their most valued possessions: automobiles.

This faith in the REPUVE program and its RFID chip extended to the business community as well. Groups such as the Mexican Association for the Automobile Industry and the Mexican Association of Automobile Dealers (AMDA) supported the program to increase car sales; it protected the domestic market from low-cost imports from the United States, colloquially known as *chatarra,* or junk heaps.[16] Already in talks with the government about restricting the importation of older vehicles, the Mexican automobile industry saw the REPUVE as a possible tool for serving its constituency. "We need a second step related to the establishment of the Public Registry of Vehicles," AMDA president José Gómez Báez announced to the press, "as an axis of control over plates for the purposes of public security, vehicular control, mechanical inspections, and pollution emissions."[17]

My interviews with Sucaro and the Velocity Motor Corporation (VMC) representatives, if mainly capturing their unhappiness with the costs of the REPUVE, also revealed support for the idea behind the program. Agustín Sandoval, from Sucaro, the car importer, echoed individual drivers in describing how increasing security over vehicles would support Sucaro's business. "It complicates things for thieves," he argued, "because those cars that are stolen the most are the new ones. So, for us, it [REPUVE] helps because last year the industry suffered thefts of new cars from carriers. They didn't even reach the dealers. Now that they have chips, it helps. Maybe just a little, but it helps. If they steal six cars, it's not as easy for you to just put plates on them because it's reported to REPUVE."

Felipe Ibarra, also with Sucaro, painted a similarly positive picture of what the REPUVE could do for car companies. He saw communications with REPUVE's Relations with Obligated Subjects Directorate as a benefit, one that helped build trust in the program and contrasted with previous automobile registries and the RENAUT, which was very much in the news at the time of our conversation. "The REPUVE was born much more solidly than the RENAVE and its predecessor," he told me, "in terms of having clearer plans for confidentiality. They took us to installations. When the program was launched, the secretary of public security took us to the Iztapalapa Airport. So, I think that from when it was born up to now, the steps have been a little slower than we would have liked, all of them, but they have functioned well." Ibarra later came to the case of the RENAUT:

> There is an example, completely opposite, that hasn't functioned, and that's the one with cell phones. At its core, the cellular registry has never stopped being a failure, it didn't work and then it didn't work. I don't think they [the cell service providers] understood it as a benefit, but only as a detriment. And it affected their market. In the case of cellular service providers, the majority of their clients are anonymous. But the problem was more about the user than the identity. They [the users] didn't have confidence in it [the RENAUT]. "I'm going to give my data, who knows who I'm giving it to?" My sister works for a telephone company, for example, I'm not going to say which, but she shared stories about clients who would buy cards and when it was time to enter their data, they would say, "Put down Mickey

Mouse, Donald, and Pluto. Put it down or I'm not buying anything." In addition, everything was cash. And in whose name? "Pancho Pantera" [an iconic advertising cartoon character who sells sugary breakfast cereal]. Things like that.

Christian Ruiz of VMC, meanwhile, told me how his company viewed working with the REPUVE as an advantage:

> We participated as a work group with the REPUVE. A REPUVE [team] came, the law was already passed, with all of their requirements: "Have this and that," "You should comply with this and that," "And this is what you're going to do." So, VMC was the pilot, because it was the first corporation to implement the process. And being the first, obviously we were able to run the first tests, the first pilots, to see how it works, where to place the sticker, how to comply with everything that they were telling us.

For VMC, then, volunteering to work with REPUVE administrators at an early stage provided the advantage of getting ahead of its competitors in learning how to implement the program efficiently.

State officials, for their part, also supported the program for its potential to secure automobility. Officials in Tamaulipas, for example, used the REPUVE database to estimate that there were at least ten thousand stolen vehicles with reports of theft circulating in their state.[18] At the REPUVE installations themselves, technicians had uncovered handfuls of stolen vehicles attempting to register for tags. As Ignacio Meza, who oversaw REPUVE Zacatecas, informed me, "We have detained various vehicles here with altered serial numbers or with reports of theft. Some situations were uncovered while checking the vehicle." Rodrigo Domínguez at the REPUVE Sonora site described a similar situation: "We have a good relationship with the people from [the] stolen vehicles [desk] at the prosecutor's office. I communicate with the prosecutor personally and he comes [here] to see what the problem is. It's not like in the movies, busting down doors and threatening people. He comes, more than anything, to see what the situation is, to interview the person, to determine whether he is guilty or not, if he has documents that show whether the vehicle was bought legally." In this sense, even absent the installation of RFID readers at transit points to instantaneously locate vehicles circulating with reports of theft, the REPUVE provided states an increased capacity to identify and capture stolen vehicles.

In addition to this, for state officials, the program lent itself to imagining how vehicles might be governed in new ways. As noted in chapter 2, Meza told me that "when Zacatecas made its first efforts [with the REPUVE], there were various states that didn't view the program in good terms, because it doesn't represent a benefit, but a cost":

> But one has to see the public security aspect and how it could benefit states. For the finance ministries, with this program, the first benefit that it gives is having certainty over the vehicle roll. To clean it, to purify it. This is one of the objectives. The other is that you can pull products from this program, such as being able to control traffic violations, to do tax operations. A few administrations ago, the state of Zacatecas started a program called Secure Transit. It was primarily a public safety program, but it also contained provisions allowing Finances to verify payment due. So, two Transit officials would go out, one person from Finances, and another person from the State Comptroller, who would serve as an observer to prevent acts of ill gains, abuse of authority, and so forth. This program allowed the Transit authorities to monitor pollution emissions, tinted windows, seatbelts, and so on. . . . If every vehicle had a chip, we could start a new program of this type. With the scanner and chips, we could collect information on payments. And this would help the operation of such a program a lot. The other would be, like I was telling you, with traffic violations. Give a reader to traffic officials with a small printer and print out traffic infractions right there. They give you your ticket, and they would send you to the Traffic or Finance Secretariat offices.

Thus, in addition to increasing vehicle security, the REPUVE could facilitate tax collection and reduce corruption among unsupervised police officers, two fundamental obstacles to governance in Mexico.

As the preceding survey and interview responses suggest, the different destinies of the REPUVE and RENAUT can be explained by the opinions held about them by ordinary people and drivers, major global corporations, and local and state governments. But these perceptions were not formed in a vacuum. Rather, as Felipe Ibarra claimed, "the REPUVE was born much more solidly than the RENAVE [National Registry of Vehicles] and its predecessor," a comment reflecting the influence of the REPUVE Obligated Subjects directorate's efforts to win his company's support of the program. The sections that follow describe additional efforts taken by administrators to build backing for the programs and overcome the resistance they engendered.

5.3 TECHNICAL RESILIENCE

In their efforts to overcome resistance, program administrators had to be willing to alter their plans in order to accommodate the forces opposing them. One simple strategy the REPUVE administrators relied on to do this was technical resilience, or allowing changes to program requirements to cultivate and maintain the participation of the *sujetos obligados*. For instance, as originally conceived, and as described in chapter 3, the application of tags to vehicles involved inputting a car owner's data into the computer database and then inserting a tag sheet into the printer to print out a tag and accompanying receipt. In effect, creating a tag generates two documents—a tag and a receipt—from an initial tag sheet. The REPUVE designers had anticipated having car producers place the receipts in vehicles, which would show car buyers that the vehicle was entered into the database and establish the REPUVE program as a presence in their minds. However, car producers balked at the complications this simple procedure would add to the production process. As Fernando Orozco with VMC explained during our tour of the company's facilities,

> At one time they [program administrators] imagined or thought that the data of the vehicle and registration would be printed out and that this paper would be placed with the owner's manual in the glove box. From behind a desk, it looks easy. It's easy to talk about. But doing it? That's the problem. When you begin to place yourself a little bit more into our production process, and everything that would be involved with having just one person going in manually and opening the door? These are all the things that would have to be changed. . . . VMC was one of those companies that said the most about this because we have everything automated, or, say, paperless.

Stressing the gap between how the world looks from "behind a desk" and "in the production process," VMC pushed to have the process for applying chips altered. And ultimately, they won.

Samuel Gallo, with the State and Federal Operations Implementation Directorate, remembered those negotiations during an interview: "We wanted those vehicles with stickers to also have a receipt that would be placed in the glove box, so that the person who bought the car would have it. It was such a problem that we finally said, 'Fine, don't include anything else. Just stick the tag on.' So, today, they just stick on the tags and they keep all of that paperwork. They destroy it. No one else keeps it." If Gallo's memory of the negotiations betrays a frustration with losing that element of the program, the willingness to alter requirements did help ensure VMC's participation.

The same flexibility defined other operations within the registry as well. The protocol for registering new vehicles specified twelve fields of data to be filled out. For vehicles already sold, Sucaro filled out this information in advance for each RFID tag it would apply. One day, however, as Agustín Sandoval told me, "we had a problem because there wasn't a technician to go to Manzanillo. So the operation had to be done in Guadalajara. But things got complicated. Normally, the ship arrives, the next day it unloads, the following day is exclusively for putting chips on, and the next day, the train or car carrier is on its way. But when the technician for whatever reason cannot make it?" This left Sucaro in a bind. "You cannot create an invoice if fields 11 and 12 [the REPUVE file and tag numbers] are missing." But at the same time, Sandoval emphasized, "there's no way we're holding up a sale, right?" In this case, Sucaro was able to resolve the problem through a simple communication with the REPUVE: "We gave notice to the REPUVE as a contingency. And with the authorization of the REPUVE, we were allowed to start the paperwork and later update fields 11 and 12."

Technical modifications also extended to the positioning of chips. Tomás Ayala, with Procedures and Citizenry at the REPUVE, described the directorate's flexibility in this regard:

> We provide the most recommended and least recommended position for applying chips. But we leave things open in an effort to be negotiable, in order to not have so much resistance to the placement of the tag. So, the most recommended zone for sticking it on is obviously behind the rearview mirror. Why? If you are sitting inside the vehicle, you don't see anything. You see it outside. Maybe it's not the most aesthetic thing. But the most important thing is the security of the vehicle and driver. But this isn't always followed. There are two companies now that are sticking the tag on the lower part of the windshield.

While the administrators believed that a tag placed on the lower windshield threatened drivers' visibility, they allowed it to be put there to reduce the automotive industry's resistance to the program.

5.4 PROVIDING QUALITY SERVICE

Another simple and inexpensive strategy available to program administrators for cultivating compliance with the REPUVE was the provision of quality service, or anticipating, listening to, and addressing the concerns of those mandated to participate in the program. For example, regular communication was central to the interactions between the REPUVE Obligated Subjects directorate and the *sujetos obligados* responsible for applying tags, as the comments from car company representatives have indicated. Julieta Salazar, with the REPUVE Obligated Subjects directorate, stressed the scale of communications with *sujetos obligados*. A "key for maintaining the program's operation," she told me, "is the help that we have provided in terms of consultations via telephone, electronic mail, and training. From 2007 to mid-2012, 173 training courses were given, with 2,015 *sujetos obligados* in attendance and a total of 4,243 users." In 2009, Salazar continued, the program's work with *sujetos obligados* focused "principally on the manufacturers, in order to be able to implement the process of applying and recording registration tags, which involved answering their questions, providing them procedures, doing checklists of the requirements they had to meet, arranging meetings with each of them to review their operation and to review how they were implementing the application and recording of tags. Accompanying them in the implementation has been key."

As VMC's Christian Ruiz commented, the REPUVE directors' demonstrations of the program provided his company an opportunity to conduct pilot tests and ensure his company's participation. Salazar expanded on this point:

> In order to optimize testing with the industry in terms of time and resources, the General Directorate of the REPUVE, together with Neology [the producer of tags], proposed taking one institution as a pilot . . . in order to guarantee the functioning of the component before making it available to all the manufacturers. Three manufacturers wanted to participate and different tests were conducted by Neology and the General Directorate of REPUVE. . . . It was important to involve the manufacturers, make them participants in what the problems were and [in] who was responsible so that we could work together.

Auto company representatives seconded this view. Felipe Ibarra, with Sucaro, explained that communication "has been the advantage. Communication between us is very close, between the director [of the Obligated Subjects division] and us, who bring the information up" the chain of command to Sucaro's executives in Asia. This close communication was critical in overcoming the automobile industry's resistance to the REPUVE. Agustín Sandoval, with Sucaro, noted that although "all of the automakers are applying the chip . . . at the beginning there was a lot of resistance on the part of the industry" because of the "operational and administrative costs" and the "problems that they had had with antecedents such as the RENAVE." Getting the industry on board "was a process that took years." In particular, the tests and presentations left a positive impression on the Sucaro representatives. "The chip itself, technologically, we view very favorably," Sandoval told me. "It has been impressive to see the wonders it can do. They did tests with other technologies. The truth is that we were pleasantly surprised with the capacity of the chip. What we want is that they implement this nationwide, with all the cars and, above all, with the imported cars from the United States, the *chocolates* [cars imported into Mexico illegally], which is another problem" for national car sellers.

REPUVE administrators saw quality service as critical in their interactions with car drivers as well. As Vicente Bautista, with the REPUVE Procedures and Citizenry Directorate, described, "We do not update the database. We coordinate the updates. Who actually does the updates are the judicial authorities in the country, including vehicle thefts and recoveries. . . . The citizens have to contact the prosecutors [to find out what happened]." However, in order to better serve drivers who contacted the program, Bautista noted that program administrators would often contact state authorities on their behalf. "Give us your information," they told callers, "and we, with the goodwill that we have, with our capacity to coordinate, if we have a relationship with the state prosecutor, we'll say, 'Hey, check whether this car isn't already recovered and hasn't been taken off the list [of stolen vehicles]' or 'Change the status please.' These are a lot of the requests that we get."

At the REPUVE registration sites where drivers have tags placed on their vehicles, providing quality service had particular value. The previous chapter noted how client interactions could quickly sour when disputes over paperwork and other matters arose. Rodrigo Domínguez, the manager of the REPUVE Sonora site, felt such conflicts could be avoided through timeliness and respect. "The average time we take in attending to a vehicle is twenty minutes. The person brings their documentation, it is entered into the system, it is validated, an inspection of the vehicle is done, all of that is a process that lasts twenty minutes," he explained. However, "there are some very isolated cases that last longer," having to do with "the third serial number or an inconsistency in the documentation or something strange." In these cases, "one of the philosophies that I have emphasized to my guys, to my technicians, is not only hitting your numbers, but also doing so in a proper way with a good face, with a good attitude. And in the end, you'll be happy. Don't get upset or upset the life of someone else. Do it the right way."

Good client service, Domínguez noted, had a special importance in a state like Sonora, where politics can infuse various aspects of daily life, including vehicle registrations. It "is something that in the long, long run will change every one of us," he told me. "How you treat people at work is how you will be able to plant a seed for saying, 'Listen, you know, this person, although I'm from the PRI, PAN, PRD [Democratic

Revolution Party], or Party X, or no party, he treated me well.' Seeds of change will be planted, seeds capable of generating change. And not directed to any party in particular, but simply toward creating a culture in which the public servant is there to serve the public and not to serve other types of interests."

"Other types of interests" within public service in Mexico include corruption, which was a primary concern for Domínguez:

> People are, let's put it this way, used to doing what they want with transactions like this, because of this cultural attitude of "it doesn't matter, this is Mexico." People may come with the idea that money solves things. Or they simply want to intimidate you in the sense of saying, "I know such and such administrator, I know or I'm a relative of such and such person." One time I had a case where the person had a bad document, we couldn't process it because the document was no good. Then he started to say, in an arrogant manner, being an asshole, "You are going to apologize to me." In this module, I am in charge of making sure, by talking with the guys, with the technicians, that this doesn't happen.

But combatting corruption and influence peddling at the REPUVE site involved more than training and leadership. In contrast to what I heard from workers in Zacatecas, Domínguez emphasized that "we are doing well in terms of structure, in terms of equipment, in terms of infrastructure." The salaries, he explained, "are very good, above average, because the conditions that were discussed, not by me but my boss, were, 'If the salaries aren't good, the workers would be given to asking [registrants] for money.'"

In a place like Sonora, where the desert climate routinely produces summer temperatures of 100 degrees Fahrenheit, the fight against corruption could also be bolstered through air conditioning. In contrast with the REPUVE module I visited in Zacatecas, which was housed in two semitrailers, one of which lacked climate control, the Sonora installation had its own building, with air conditioning (fig. 23). Domínguez added

> In addition to the question of salaries, the fact of having good buildings, the fact of having adequate equipment, having the help and support of the bosses in charge of the program, [this] helps make the workplace a pleasant environment. This helps, in some way, to clean up some forms of corruption. Not having people say, 'I'm giving you this for some sodas,' or 'Here, to speed things up,' or all of these types of things. For a technician or data specialist, this would mean losing that. In this region, to be in an enclosed, air-conditioned building, I'm going to be careful that my job isn't terminated, so that I'm not selling popsicles out on the streets.

FIGURE 23. REPUVE registration site in Sonora. Photo by Keith Guzik, © 2015.

The REPUVE building in Sonora also included a waiting area, where registrants could comfortably wait while having their vehicles inspected. In addition to client service, the facilities helped create a separation between drivers and workers, which deterred influence peddling. "In the inspection area," Domínguez finished, "only authorized people are allowed. And [with] any visit that occurs to that inspection area, an authorized person from REPUVE has to, let's say, babysit the visit."

5.5 STICKS AND CARROTS

Sujetos obligados who are already complying with program requirements and drivers who voluntarily register their vehicles present the least resistance to the REPUVE program. For states attempting to implement the program, the real obstacle was getting drivers to register their cars in the first place. To this end, a perhaps reflexive response for administrators was to threaten people for their insolence. In Zacatecas, where few drivers were enrolling, REPUVE employees told people passing by the registration site that the program was "free and voluntary now" but would be "mandatory and carry a charge" later. The logic was simple enough—act now or risk losing money later.

This strategy reflected a broader belief among the team that the threat of punishment could bring people to the program. Ignacio Meza, who oversaw the site, explained that "right now, we are doing things on a voluntary basis, it's an invitation to the citizens, and they come voluntarily. . . . The federal government gave us a timeline of two years to get to 100 percent of the vehicle roll. Crunching the numbers, I don't think any state is going to reach 100 percent. But what we can achieve is not having the program be voluntary at the end of these two years. The traffic rules have to be modified to require vehicle owners to install the chip." Meza continued,

> With this modification to the traffic code, the citizen would be required, in some agreeable way, to put it like that, to come and register his vehicle. What would this look like? If you are driving your vehicle and the requirement exists, the police are going to detain you and give you a first warning, like, 'Go get your chip' and give you ten working days, or two weeks. If these two weeks go by, and you haven't come to complete the transaction, they are going to stop you again and give you three days to get it done. And if on the third time you still haven't done it, then the car is going to be detained and brought to the tow yard, and there they're going to install the chip once you present your documentation. This is the way to tighten things up.

Other states wrestled with similar issues. In Michoacán, for instance, Armando Ballinas Mayes, secretary of the State Council on Public Security, noted that the slow response to the REPUVE reflected "a culture of negligence by car owners in the state." To make the program work, he contended, it would have to be obligatory: "If we leave it to people's fancy, we run the risk that no one comes. We are looking at the law and the traffic code to make this registry obligatory and for the motorists to carry their stickers on their vehicles together with their registrations and driver's licenses."[19]

The legal status of the REPUVE as voluntary and free is interesting to consider. The federal REPUVE law clearly states that the "the registration of vehicles in the Registry is obligatory." It also notes that "the registration of vehicles, the presentation of notices, and consultations of the Registry will be free." Nevertheless, the federalist structure of the Mexican state provides states leeway in their adoption of federal laws.

In any case, as a tactic to strong-arm drivers to register their vehicles, the threat of future costs and obligations produced mixed results.

The ploy did not result in a noticeable increase in demand at the REPUVE site in Zacatecas. Perhaps the threat, made during the summer months, failed to strike fear. Only one driver I spoke with said that he was registering his vehicle in order "not to wait until the last minute and get stuck in lines, like we [in Mexico] always do," which was a primary complaint of both workers at the site and the Michoacán official. Drivers by and large stayed away. However, this same threat resulted in the rush on Ciudad Juárez's REPUVE installation, noted at the beginning of the chapter.

The stick of threatening people with fines and fees in the future can also be seen as giving people a carrot in the present: financial savings. In this vein, other states in Mexico sought to increase favorable reception of the program by offering incentives. In San Luis Potosí, where registration in the REPUVE database was to be tied to obtaining vehicle license plates and registrations, the state legislature voted to delay the REPUVE registration requirement for a year in order to give drivers more time to register with the program voluntarily.[20] In Zacatecas, after threats failed to have an effect, the finance secretary agreed to incentivize participation by offering drivers a 5 percent discount on their *tenencia* excise taxes if they registered.[21] In this way, states tried to offer drivers benefits in the form of additional time and financial savings to win their participation.

5.6 BUILDING CRITICAL MASS

REPUVE administrators tasked with enlisting the participation of states reluctant to cooperate with the program faced a distinct challenge. In contrast to the strategies employed with *sujetos obligados* or drivers, REPUVE's State and Federal Operations directorate could not threaten states into participating, since it lacked that authority under Mexico's federalist system. In addition, in summer 2011, the REPUVE leadership faced an upcoming presidential election that most believed would bring the PRI, and its candidate, Enrique Peña Nieto, back to power. As noted in the last chapter, the expected transition raised concerns that the entire program could be lost. Looking toward the future, Samuel Gallo, with the State and Federal Operations directorate, told me that "2013 is critical for us. The new boss arrives and there could be some decisions that could really, all of the sudden, throw this program in which they [the federal government] have invested five hundred million pesos in six years down the drain. And programs have disappeared in the past. . . . For me, it's a concern because, if we haven't shored up an increase in states that are implementing the program, practically speaking, in the next elections, they [the new administration] would say this program has presented more problems than solutions."

Faced with this time constraint and lack of progress, Gallo and the rest of the directorate saw the REPUVE's survival as tied to building a critical mass of participation. As Gallo explained it,

> Our objective in the area of operations is to advance as much as possible into the middle of the next year [2012]. . . . Right now, at the moment, we have fourteen states and are calculating to see if we can close with twenty states operating. We are talking 60 percent of the country working with the program. . . . If some states get on board, and some others have already invested in it, it's going to be difficult if there is a change of administration at the federal level to throw this away. Why? Because some states are going to jump up and say, "Hey, I already invested so many millions of pesos, so I can't throw this in the trash. Besides that, it's helping me do this, that, and the other thing."

If program could take root and begin to bear fruit, REPUVE administrators could better ensure its continuity into the future.

Reaching a critical mass of state participation involved some of the same tactics used to gain *sujetos obligados'* and drivers' support. Communication was again critical. Addressing this point, Gallo explained, "We have been able to talk with governors, with governmental secretaries. We have put ourselves where we don't belong in order to get the project going. . . . We have national forums, which prosecutors and governors attend."

During summer 2012, I attended one of these national forums, convened by REPUVE's State and Federal Operations directorate. Representatives from seventeen states attended the forum, held at the Convention Center in Tlaxcala, one of the states championed by the directorate as exemplary in its progress with the registry. Seated around a horseshoe-shaped table, the representatives listened to presentations from REPUVE Veracruz and REPUVE Sonora, which had successfully implemented the tagging program. Following the presentations, a question and answer session, and a finely prepared lunch, the representatives were led outside to the Convention Center's parking lot, where a trailer similar to the one used in Zacatecas was set up to give a demonstration of the process for applying tags to vehicles.

FIGURE 24. Video screen displaying REPUVE camera feeds at a REPUVE national assembly in Tlaxcala. Photo by Keith Guzik, © 2015.

As a recruitment tool, the forum emphasized the security situation facing the country. "In Mexico, in the last few years, we have had a surge of insecurity in all states," the presider emphasized. "In 2009, there were, if I remember correctly, 67,000 vehicle thefts, and in 2011 we had almost 80,000. And these are insured vehicles. There are many that are not insured and are not in the statistics . . . and this leads to a lack of citizen confidence in authorities. Therefore, we have to join forces to stop it [the insecurity]. We have to communicate. We have to use new techniques. We have to try to combat it head on and prevent these crimes."

To demonstrate the REPUVE's potential to fight crime, the forum organizers arranged for a car that had been entered into the registry as stolen to pass through a toll installed with a camera and RFID reader. Through a video link set up at the toll area (fig. 24), attendees watched as the vehicle was detected as stolen and police cars were deployed to stop it. The demonstration provided a clear vision of the secure future to which the REPUVE could carry Mexico.

If the potential of the REPUVE to strengthen governance was something conference attendees had heard before, the Tlaxcala meeting also showed that it could done. The program was viable, as REPUVE Veracruz and REPUVE Sonora demonstrated. The Operations Implementation team told attendees, "The fundamental objective of this event is to demonstrate and in some way make visible that the program is a reality, that the program functions, that the program is being implemented in our states. We have fourteen states at the national level that are already working in this process, and the reality is that this is a benefit for the citizens of your states. . . . We have a database that contains all of the work" from the states, vehicle manufacturers, and customs offices. "In total, this is almost thirty-three million vehicles. All of these registrations have the same quality and same authentication of data. Of these thirty-three million vehicles, no [vehicle identification] numbers are repeated. There is not a single repetition in the Public Registry of Vehicles. All have been verified, which allows us to say to you that there are no false entries. The database has integrity." The REPUVE representatives continued on this theme: "What motivates us to be here and speak to you about how the program functions is that the program is already working, that we have really been meeting our goals. . . . We have practically 50 percent of the states at the national level applying registration tags."

Such talk of progress conveys a program with momentum. But the presentation in Tlaxcala was also calculated to pressure states to join. On another occasion, Samuel Gallo, who had helped put the Tlaxcala event together, described to me the national forums in general: "Graphs are presented and then you can say, 'You haven't completed your task. You haven't completed your work.' In this country, what hurts the most is when you tell me in front of everyone that I didn't do my things. It works like this. 'When in Rome, do as the Romans do,' as the saying goes. And if you don't like it? Go do your job. Because I'm going to be here every three months in the meetings that we have with executive secretaries, with prosecutors, with governors, and the statistics will be presented." By demonstrating other states' progress to key decision makers, the REPUVE planners felt better able to push other states to participate.

Taking this idea of shaming further, the REPUVE State and Federal Operations directorate also tried to reach a critical mass of state participation through publicity. This strategy is counterintuitive, since states rather than drivers are being targeted. But as Gallo explained, "We are trying to implement a strategy that consists of a webpage—we are going to have it ready soon—in which we will report the situation of every state in the REPUVE. So, there will be states that have 50 percent implementation, or states that have 80 percent, or states that have 0 percent. This [web]page is for citizens' consultation. People are going to enter [the website] and they're going to see that the REPUVE in some states is already applied and in other states no." This public reporting, Gallo believed, would lead citizens to pressure their state governments to join the program. "If mine isn't there," he imagined people thinking, "I'm going to apply social pressure, so that the government carries out its job. Because if not, I'm not going to pay taxes to you. I'm going to pay them to a neighboring state that is already applying chips and get plates there. And I don't give a damn. . . . If my government isn't concerned about security or protecting my interests, then I as a citizen don't want you here. Then, we [the citizens] are going to start a blog, we are going to post questions." In this scenario, Gallo not only imagines the program as one in demand among ordinary Mexicans, which my survey data partially support, but also calculates that popular dissatisfaction with authorities could be turned to the program's favor.

5.7 ALLOWING LEVIES

The ability of the REPUVE leadership to force states to participate in the vehicle registry was, as mentioned previously, legally limited by Mexico's federalist system of government. Similarly, the program's ability to incentivize states' participation was limited by scarce resources. As the Operations Implementation team told representatives at the Tlaxcala conference, "We have to carry out the program in a mindful way. However much money they give us, it's not going to be a billion pesos, or two billion pesos. So we have to do origami, work in stages, work with concrete plans, and strategies." In lieu of a good stick or large enough carrot with which to impel state authorities to adhere tags, the directorate had to consider other options.

During summer 2011, news reports circulated that customs offices at Mexico's northern border would begin charging $54 for a "new security hologram for automobiles" imported into the country.[22] The fee would be charged "for the inspection, physical review, sticker, and subscription in the REPUVE."[23] "Importers of vehicles would have to pay Banjército," the National Bank of the Army, Air Force, and Navy, which is, as its name suggests, a banking institution for members of the Mexican armed forces.[24]

Smaller car importers in Mexico, who make their living transporting vehicles from the United States for sale in Mexico, and residents of the border areas, who were accustomed to crossing the border to purchase cheaper vehicles in the States, were furious with the new regulation. Raúl Quintanilla, an importer, complained that Mexican customs agents had "pulled this out of their sleeve" and that the measure would "harm us even further. This sticker is supposedly for registration in the REPUVE. . . . This is going to affect the number of imports that we do. We are already having problems doing forty cars a day. The nearly nonexistent importation of cars will be severely affected by this new charge. Importing cars is out of reach."[25]

Business associations, such as the local branch of the National Chamber of Commerce (CANACO) and the Independent Union of Car and Truck Sellers (UIVAC), added their voices to the chorus of complaints about the import charge. Besides the new import fee, these organizations bemoaned the poor quality of service at the customs offices. Emilio Girón, president of the local CANACO chapter, complained that "with this, the costs are higher and the operations slower. We know that [before] they could complete eight thousand operations per month and finish each one in a minute. Now it's going to be thirty minutes to put on a sticker. Banjército doesn't have enough people, and they don't do their work adequately, and the quality of the sticker is awful."[26] Alfonso Delgado, president of UIVAC, denounced the fee as "an unjust charge because the [REPUVE] law specifies it [the tag] as free." In addition, Banjército was collecting the fee in US dollars and giving receipts without an official stamp or a breakdown of the value-added tax. Finally, more than a month after paying for the inspection of vehicles, Delgado said, cars were still appearing on the REPUVE online portal as "not registered."[27]

Mexican customs offices' charging for registration in the REPUVE illustrates another tactic by which the REPUVE State and Federal Operations directorate won the backing of state actors: by accommodating their power to tax. Allowing customs offices to collect fees for the REPUVE is interesting because, as Delgado correctly noted in his criticism of the fees and as has been noted before, the REPUVE law says that "the registration of vehicles, presentation of notices, and consultations in the Registry will be free." What's more, this provision of the law directly responded to the controversy surrounding REPUVE's failed predecessor, the RENAVE, which charged a subscription fee for its (absent) security services.

But within the national REPUVE offices, the controversy over the customs fees was viewed less critically. When I asked Samuel Gallo about the matter and the seeming contradiction between the law's calling for the REPUVE to be free and the actions of the customs offices, he responded, "The law tells us that the printing, recording, and application of the tag cannot have a cost. It's free. So we cannot charge for this process because the law says so. However, some states are taking advantage of the fact that the physical inspection done to vehicles and the inspection of documents can be charged for." Allowing for the apparent contradiction, Gallo conceded,

> At the end of the day, there are funds and there are expenses. So, in some way, this [inspection fee] offsets that part. They [the states] imagined covering costs with the FASP [Support Funds for Public Security], from that support fund from the federal government. But it was insufficient. The projects are large. There are many other priorities as well, the certification of police, the program to create a uniform police force. There is a large number of projects for which there isn't enough [funding]. In this sense, the money is lacking. So some states have started charging for it [the tag]. There is a clear line between the application of the chip and the physical inspection.

By distinguishing between the actual application of tags and the procedures needed to authorize their application, the REPUVE directorate was able to mark off a sphere of activity in which fees could be levied while remaining consistent with the letter, if not the spirit, of the law.

Speaking specifically to the situation with the customs offices, Gallo said,

> When we went to the Finance Secretariat, we said, "Look, by law it's your job to give me the information on imports, both permanent and temporary." . . . We had been negotiating with them for seven years until last year, when the decision was made to really begin with the program. Their argument was that, "I don't have the ability to carry out the process that you are asking for. I don't have the capacity or the space or the infrastructure. I don't have the personnel to do it." With such a tone, it was very complicated. The Support Funds for Public Security are only for the states. . . . They told us, "Well, there's no money, I don't have the people, I don't have the capacity."

In this logjam, the resources and operational capacity of Banjército offered a solution. "Banjército had already been operating a program for years," Gallo told me, "putting stickers on vehicles imported temporarily. In this way, the possibility arose that it could be Banjército that would assist the Finance Secretariat. An agreement was reached. Banjército said, 'I will help you, but I'm going to charge for the service. I'm going to contract people, buy equipment, put up installations, create an infrastructure in order to be able to meet this requirement that you can't.' This is practically what they said to the Finance Secretariat."

While this presented a solution that would get the customs offices on board with the program, the potential fallout from the inspection fees created a predicament. In response, Gallo explained, "Our position was, 'I don't have a problem with it so long as you don't say it's REPUVE charging.' Because the law says that it's free. So, 'You have to be very clear with everything. What you're charging for is the process of the physical inspection and registration of the vehicle. You're not charging for the subscription to the Public Registry of Vehicles.' For me, it's totally transparent so long as it's not appearing [as a fee] from REPUVE. The moment it appears from REPUVE, then we're going to stop it."

Asked whether the Banjército fee made the REPUVE a burden for people, Gallo continued, "At the societal level, it's relative. Whoever is going to buy a vehicle on the other side is going to get it for $3,000 or $4,000 at most. Add to that whatever it costs to import the vehicle, to pay the import taxes and the fees, to legalize the vehicle, and now REPUVE." In other words, those who cross the border to buy a vehicle would still be coming out ahead. Thus, importers really did not have a strong leg to stand on when complaining about the charges. Further, it is worth remembering that both the Mexican Association for the Automobile Industry and the Association of Mexican Automobile Dealers had supported the REPUVE precisely for its potential to exert tighter control over the import of cheap vehicles from the United States, which hurt domestic sales. In this sense, allowing state entities to levy taxes, an improvisational tactic to reach a critical mass of governmental support for the REPUVE, fulfilled some of the program's promise for national business organizations.

5.8 PROGRAMMATIC PIGGYBACKING

Up to this point, this chapter has focused on the strategies used by state authorities to overcome resistance to the REPUVE. Turning attention to the CEDI and RENAUT, however, offers additional insights into the tactics program administrators used to counter resistance. To return to the challenges confronting the Citizen Identity Card, from the time of its launch, the Calderón administration faced strong opposition from the Federal Electoral Institute, which feared that the new identity card would be confused with the country's voter card. Attempting to find a way out of the impasse, in late 2010–early 2011, Francisco Blake

Mora, the interior secretary, convened talks with members of the IFE to resolve their differences. Blake addressed the IFE's mistrust of the administration by making Manuel López Bernal, ex-executive secretary of the IFE, the director of the National Population Registry, the organization responsible for implementing the CEDI. López Bernal resigned the following year. But Blake repeated his maneuvering by replacing him with Alberto Alonso y Coria, who had directed the Federal Registry of Voters for nine years and was the most vocal opponent of the CEDI.

In addition, Blake announced in January 2011 that the CEDI would be distributed to the entire Mexican population as planned within five years, but it would be restricted for the time being to those not currently in the voter registry, that is, Mexicans under the age of eighteen.[28] Leonardo Valdés, president of the IFE, denounced the government's decision. "The decree is undoubtedly legal," he said, "but inopportune electorally from my point of view. . . . For over a year we have been saying that the creation of the identity card would negatively affect the two pillars of our political system: the voter roll and the voter card."[29] Interestingly, though, the issuance of identity cards to minors was not an immediate point of conflict.

Blake's strategy to begin the CEDI with Mexicans under the age of eighteen contains an element of building critical mass. By beginning with only a segment of the population, the administration could hope to gradually grow the program while the conflict with the IFE resolved itself. And if the conflict could not be resolved, the CEDI would already be in public circulation.

But the move to fashion the CEDI into a Identity Card for Minors had other dimensions relevant to this discussion of power and resistance. By recasting the identity card as one for minors, the government was able to reframe the technological object. If the CEDI was imagined as a technology that would integrate other forms of identification in Mexico, thereby simplifying people's daily transactions, the Identity Card for Minors would serve as a technology to protect children from kidnappings and trafficking, as well as to facilitate their registering for school, accessing health care, or obtaining a passport. This reframing eluded the point of resistance with the IFE—if the CEDI was not being issued to voters, it could not threaten democracy. But more than this, it provided the CEDI a purpose that few could disagree with: the protection of children. As a result, the government's actions legitimized the technology in a way that the original identity card failed to do. On January 13, 2011, only days after Blake's announcement, Susan Sottoli, a representative with UNICEF Mexico, expressed her support for the Identity Card for Minors, qualifying it as a fundamental measure for guaranteeing the right to identity for children, for exercising their full rights to education and health programs, and for protecting them from trafficking.[30]

Beyond its legitimation as a security tool for children, the CEDI was also aided by the institutional associations accompanying childhood. The card was to be distributed at Mexico's public schools, an institution that enjoys a high level of support in Mexico.[31] As the events highlighted at the beginning chapter 2 illustrate, the launching of the Personal Identity Card for Minors made conspicuous use of public schools.[32] In sum, to overcome the resistance of the IFE to the CEDI, the interior secretary of the Calderón administration sought to *piggyback* the questionable program onto an institutional and ideological framework that already enjoyed legitimacy with the Mexican people.

Institutional piggybacking occurred with the REPUVE as well. Never was the program's RFID tag imagined as having an application outside providing "legal certainty" through the registration, inspection, and regulation of vehicles. However, in the state of Sonora, the tag came to possess another purpose: a technological solution for the state's tolling system, referred to as the Libre Tránsito (Free Transit) program. Under Libre Tránsito, residents in certain areas of southern Sonora are able to travel on the federal highway free of charge. Three toll plazas—the Don, Fundición, and Esperanza stations—are covered by the program. Given the toll of sixty-five pesos (about six dollars) charged at each plaza, the program provides significant savings to motorists.

The REPUVE's refashioning into a tolling solution has a peculiar history, as Carlos Aguilar, an employee with REPUVE Sonora first described to me. Article 11 of the Mexican Constitution stipulates that "every person has the right to enter the Republic, exit, travel its territory, and move residence, without needing a security card, passport, letter of safe passage, or other similar requirements." This has been interpreted to mean that a toll road cannot exist where there is no other alternative for road travel. In most parts of the country, this does not pose a problem. The federal toll roads, which are generally high quality and well-maintained, have been constructed alongside older roads. Thus, "free transit" is always available for those unable to afford the costs of traveling on the federal highways. In Sonora, however, as Aguilar explained to me, "they constructed the toll road over the free road. So, if one travels, there is only the tollway. You always have to pay."

Winning the right to travel freely on the federal highway became a cause célèbre in southern Sonora, attracting residents' ire and shaping politicians' agendas. Eventually, through negotiations between the governor's office, state representatives, and the federal Communications and Transportation Secretariat, an agreement was reached in 2011 whereby the REPUVE tag would be used to provide free passage through the three tolls to more than 700,000 residents.[33] Thus, finished Aguilar, "if you resister in the REPUVE, and if you meet the requirements of Federal Roads and Bridges [the agency in charge of tolling on federal highways], which constructed the highways, they grant free passage. And when you come with your vehicle to the toll plaza, the antenna at the toll plaza recognizes the vehicle, and if you are registered as a resident with us, the antenna lets you pass without paying a peso" (fig. 25).

As might be expected, free travel provided a strong incentive to residents to get tags on their vehicles. When I spoke with drivers at one of the two REPUVE sites then operating in Sonora, both in the southern part of the state, all of the drivers mentioned the potential savings as their reason for registering:

> The program gives you the benefit of not having to pay the toll. There are three tollbooths around Obregón. You save money.

> The program works well. Why? Because, before, passing through the tolls exacted a price. It's 60 pesos here in Fundición, 60 on the way back—120 pesos. And if I travel to Guaymas, the same.

> The tollbooth is expensive. And when I moved here, gasoline got more expensive as well. You have to invest so much just to get to a place that's an hour away, I pay fifty pesos in gasoline and seventy in tolls. So it's expensive for me.

Reflecting on the REPUVE program, Rodrigo Domínguez, manager at the Sonora site, opined, "It's important to correct the errors of the past. I see this program, speaking specifically in the state of Sonora, as a solution to a very old problem that is about twenty years old, since the tollbooth was installed in Fundición. For more than twenty years since they installed them, the cities of Navojoa and Obregón were enclosed by these toll plazas and international highways. It was a fairly simple and viable solution to guarantee the constitutional right of the residents of the south of Sonora to travel on the federal highways."

Similar to the Citizen Identity Card, *piggybacking* the REPUVE onto the existing tolling infrastructure in Sonora legitimized the program by framing it within a larger public issue that already had salience for the local population. In addition, as with the Identity Card for

FIGURE 25. REPUVE toll lanes in Sonora. Photo by Keith Guzik, © 2015.

Minors, which made use of Mexico's schools, the tolling infrastructure in Sonora ensured that many of the physical requirements for the REPUVE program, such as gates and arches from which to position chip readers, were already in place. The result of this piggybacking was a more successful and seamless implementation of the REPUVE than occurred elsewhere.

5.9 CIRCUMVENTING THE STATE

The strategy of piggybacking, or enfolding surveillance programs and technologies into existing state infrastructures, assumes the existence of infrastructures to tie into. Sometimes, however, these are not available. Mexico's National Registry of Mobile Telephone Users reached a impasse when it could not be determined who, whether the Federal Commission of Telecommunications (COFETEL) or mobile service providers, had the responsibility for verifying the phone numbers registered with the government. Eventually, as frustration with the phone registry continued, the Senate voted in April 2011 to terminate the registry,[34] and the Interior Secretariat destroyed the data it had collected in April 2013.[35]

This did not exhaust the government's attempts to govern mobile telephony, however, the original reason RENAUT was launched.

Already in May 2010, in the midst of the controversy over the requirement to register cellular lines with the government, two mobile service providers, Telcel and Movistar, announced that they would begin deactivating phones that their clients reported as stolen. But in contrast to the RENAUT, which sought to gain a grasp of mobile telephony via the phone numbers and biometric data of users, the service providers would instead disable units through their IMEI (International Mobile Equipment Identity) codes, the unique fifteen-digit identification numbers, similar to an automobile's vehicle identification number, which identify devices on mobile networks.[36] Later that year, in December 2010, this private initiative led to the Agreement to Avoid Reusing Stolen Cellular Telephones, an accord between the Citizens' Council for Public Security and Justice Provision in Mexico City and mobile companies to expand the initiative to include additional providers. Under the plan, users would not need to make an official report of a stolen phone, but simply report the theft to their service provider so that the device's IMEI could be blacklisted and the device blocked. Jorge Arreola, director of compliance with Telefónica México, feted the agreement as a collaborative effort between the Mexican citizenry and private sector.[37] And by August of the following year, Telcel and Movistar were reporting a monthly deactivation rate of twelve hundred units.[38]

While the private sector was moving to provide an alternative means for controlling stolen mobile phones, others were pressing the federal government to consider alternatives to the RENAUT. In April 2011, Alejandro Martí, the businessman who founded the organization Mexico SOS following the kidnapping and murder of his son, began pushing for legislation that would require service providers to permanently block units reported as stolen and require units to have GPS capabilities and "panic buttons."[39] In March of the following year, Congress passed what came to be known at the Geolocation Law, which requires mobile service providers in emergency situations—kidnappings, extortions, or medical emergencies—to provide law enforcement with the location of the person's telephone without a judicial warrant.[40]

The Geolocation Law evoked anger from many of the same groups that expressed concerns about the privacy and security implications of the RENAUT. The Electronic Frontier Foundation, for instance, criticized the Geolocation Law as having "a significant potential for abuse" and demanded that the Mexican government "be more sensible to the fact that mobile service companies today record detailed imprints of our daily lives."[41] But for the purposes of this work, the actions of service providers and the Mexican legislature demonstrate another way in which authorities improvise to overcome resistance to their plans. In the case of regulating mobile telephony, Mexican authorities and stakeholders attempted to push past the resistance that led to the failure of the RENAUT by trying to contravene the state altogether. Unable to create a state infrastructure that would be able to register, inspect, and regulate mobile telephony as set out in the RENAUT law, authorities simply legislated the burden of regulating phones onto private service providers and the private infrastructures through which mobile communication occurs.

5.10 STATECRAFT

The preceding pages have sketched out some of the strategies that authorities in Mexico have pursued to overcome resistance to the Public Registry of Vehicles, Citizen Identity Card, and National Registry of Mobile Telephone Users. Confronted with corporate actors skeptical of the costs of complying with new regulations, governmental authorities maintained close and regular *communications* to increase understanding of the programs and permitted *technical modifications* to their operation to make them less onerous. To counter individuals' resistance, authorities aimed to achieve good relations with users through quality *client service* and experimented with combinations of *threats* and *incentives* to bring new users into the fold. To answer resistance from the political structure tasked with carrying out the programs, authorities required more imaginative tactics. Facing political turnover that could threaten program continuity, authorities sought a *critical mass* to provide a beachhead of programmatic progress able to withstand the ebbs and flows of democratic governance in Mexico. To achieve that critical mass, authorities again turned to *incentives,* such as the promise of security, but also to the authorization of *levies* that increased the appeal of programs to state actors. Or they *piggybacked* unpopular programs onto existing state infrastructures that already enjoyed legitimacy. Or, most extreme, they *contravened the state* altogether, relying on the expertise and infrastructure of the private sector to implement operations that might otherwise fail.

These strategies all share the quality of being improvised responses hatched on the fly to counter resistance. To surmount the inability of the COFETEL to successfully implement the RENAUT, Mexican authorities went outside the state to follow mobile phones. To work itself out of the conflict with the IFE regarding the CEDI, the Calderón administration redirected the identity card through Mexico's schools and rebranded it as a measure for youth. To get the customs offices in Mexico to begin applying chips to vehicles, the REPUVE allowed the Banjército to charge for inspections. None of these actions had been planned in advance. Like the agronomists of the 1930 Agrarian Census who had to resort to exchanging water for data from *campesinos*,[42] these measures were embraced by authorities to pursue control over collective agencies that persisted in their resistance.

These findings contribute to thinking about the state in Mexico and state formation in general. Research on Mexico has tended to emphasize the role of capital and coercion[43] and cultural hegemony[44] in the creation of the state. National state formation has varied across Mexican history in relation to the accumulation of coercion and capital in regional centers and the concentration of cultural power in the Catholic Church, processes that have worked against the centralization of state power. Against this backdrop, the national state can be understood to fluctuate historically along the axes of its strength (strong/weak) and its relation to dominant class interests (independent/dependent). The Porfiriato was primarily a weak, small state captured by the capitalist class, while the postrevolutionary regime of Lázaro Cárdenas was a large, strong state autonomous of capitalist class interests. The abandonment of the revolutionary project by the PRI in the 1980s and the party's embrace of neoliberal political economy, meanwhile, reduced the size of the state and brought it back into the service of the country's capitalist class, whose economic interests now operate at the global level.[45] And it is this phase of state formation that has occasioned the growth of organized crime in the country.

Borrowing from recent work on the role of science and technology in the "co-production"[46] or "composition"[47] of the state, this study has examined the federal government's adoption of surveillance technologies as an effort to reform the state by remaking its coercive capacity through *prohesion*. If the previous chapter revealed resistance to this project of state reformation, then the present chapter describes a set of activities—improvisational tactics launched by program administrators, state officials, and ordinary employees in their meetings with company representatives, conferences with state leaders, and interactions with everyday drivers—that helps decide the fate of state reformation.

That authorities and state workers in Mexico improvise to meet their goals should not surprise us. For all the explanatory weight that Charles

Tilly placed on "capital and coercion" in the emergence of the national state, he also noted that "rarely did Europe's princes have in mind a precise model of the sort of state they were producing" and that "the principal components of national states [are] . . . usually formed as more or less inadvertent by-products of efforts to carry out more immediate tasks, especially the creation and support of an armed force."[48] This inventive element of state formation has been noted within Mexico as well, as scholars have noted that "the stable, centralized rule of the PRI . . . rested on a series of tacit deals and trade-offs."[49] In this sense, the Mexican state is a "negotiated state."[50]

These processes of negotiation and improvisation are important to consider, not only because they are so central to state formation, but also because they draw attention to a human element in state formation beyond capital, coercion, culture, and technology. That is, REPUVE administrators had to have the ability to negotiate with Banjército to overcome barriers to customs offices' participation in the program. CEDI planners had to have the wisdom to know that programs tied to Mexico's schools held greater legitimacy with the populace. And federal legislators, with no state-based options for controlling cell phones, had to have the flexibility to abandon a failed plan and redirect the governance of mobile telephony through the private sector. In brief, the histories of the REPUVE, CEDI, and RENAUT demonstrate that reforming the state through surveillance technologies requires a good dose of managerial tact from program planners and administrators.

To give proper emphasis to these skill-based improvisational activities that are integral to the fate of state projects, it seems appropriate to name them formally. To this end, we might repurpose the word "statecraft." Statecraft is a term that has traditionally denoted international diplomacy. But the term itself, in accentuating craft, or work requiring particular skill, captures well the practical, hands-on dimensions of state formation. Thus, to apply that term here, it could be said that Mexico's experiences in trying to adopt surveillance technologies to fight insecurity reveal the importance of statecraft in state formation.

Identifying the role of statecraft in state formation is not to argue that the fate of the state is decided by statecraft alone. The notion of art or improvisation inherent in the term indicates the lack of control that those practicing statecraft have as a condition of their work. In the cases of the RENAUT, CEDI, and REPUVE, the availability of resources, the extant organization of the state, the positions of major businesses relative to the programs, the opinions of the public, the technical designs of the programs, and so forth were all forces that conditioned program outcomes. But what statecraft underscores is how state administrators and planners are still able to problem-solve and construct the state when confronted with such obstacles.

The improvisational nature of statecraft lends state formation an unpredictable character. That is, if tactics such as those outlined in this chapter give authorities a hold on automobility, mobile telephony, and personal identification, they are not the holds that were planned. In Sonora, where the application of chips has been successful, the REPUVE program functions and is understood, as the quotations at the start of this chapter indicate, as a technology allowing local residents to pass freely through federal tollbooths rather than a program for locating suspicious vehicles or ensuring "legal certainty." In Baja California, Colima, Chiapas, Guanajuato, Jalisco, and Nuevo León, personal identity cards have been distributed to Mexicans, but only those under the age of eighteen, and are thus understood as tools to facilitate youth access to schools and public services rather than tools for all Mexicans to integrate and secure various forms of personal identification. Mobile telephones, meanwhile, while intended to be secured by the government's Interior Secretariat, end up being the responsibility of private actors considered better situated and skilled to handle them.

Uncertainty does not mean that statecraft is random. In the modifications made to programs, state administrators take advantage of extant ideas and arrangements of the state that are already operational. This is clearest in the practice of programmatic piggybacking, where institutions that enjoy popular legitimacy (public schools) and material infrastructures that are in use (tollbooths) are employed to advance fledgling programs. But it can also be seen in encouraging state agencies' authority to tax (the case of Banjército) or accommodating the legal claims of citizens (Free Transit in Sonora). So, then, if statecraft entails uncertainty, it is a patterned uncertainty that gravitates toward past and present state forms.

Saying this, past performance does not guarantee future results. And whichever hold that statecraft allows authorities over collective agencies in society cannot be permanent. For instance, the Calderón administration's attempt to circumvent the state by deputizing cellular service providers to track mobile devices was met with strong public opposition[51] and challenged by the National Commission on Human Rights before the Supreme Court. Although the law was ruled constitutional,[52] once Enrique Peña Nieto assumed the presidency in 2013, his administration sought to make its own imprint on the governance of telecommunications by pursuing the Telecom Law, a series of reforms widely criticized as a threat to net neutrality.[53]

The Personal Identity Card for Minors, meanwhile, was met with skepticism and confusion by those it was intended to benefit: the parents of Mexican schoolchildren. "How do I know that they won't misuse our children's data?" asked Verónica López, mother of one child whose school was registering students. "A lady said that this would help a lot because something like this is already done in Europe. But, for us, it's new."[54] For its part, the lower house of the Mexican Congress sued to halt the program on the grounds that collecting iris scans and fingerprints from minors represented a violation of privacy.[55] The Supreme Court upheld the program's constitutionality, but the new administration eventually halted distribution of Identity Cards for Minors and launched an investigation of the program's use of state funds,[56] all while reemphasizing the federal government's intention to realize a national identity card for all Mexicans.[57] Meanwhile, the Federal Electoral Institute, in its attempt to maintain its voter card as Mexico's primary form of personal identification, went forward with a redesign of the card that would meet the security requirements demanded by the federal government.[58]

Finally, the Mexican government's attempts to implement the REPUVE at border crossings by allowing Banjército and customs officers to charge for inspections led to lawsuits sponsored by the PAN.[59] The lawsuits resulted in Banjército's suspending the program at those sites, leaving vehicles crossing into the country outside the registry.[60] Thus, despite the efforts of authorities to improvise different ways to implement these surveillance technologies, the government's hold on mobile telephony, personal identification, and automobility remains tenuous.

These considerations, like the histories shared in the second chapter, suggest a metamorphic and temporal quality to state formation. The tactics that succeed in gaining a hold over collective agencies are defined and transformed by the resistance they encounter. And the success of these tactics can be fleeting as new forces push back against state arrangements to control them. New tactics must then be adopted to provide a grip over collective agencies anew.

What results from the state's adoption of surveillance technologies is not prohesion, then, whereby the state attaches itself to the materiality of collective agency en masse and brings its diverse assortment of governmental agencies together into a single network, but a further hybridization of strategies and tactics that mocks the unifying impulse of prohesion. This does not mean that prohesion has failed to take root in Mexico. The REPUVE database exists and is used by public authorities and private actors to provide "legal certainty" about the provenance of certain vehicles. Vehicles circulating in Sonora, crossing over the Mexican-US border, operating in Zacatecas, Tlaxcala, and other states, and purchased from new-car dealers have tags attached to them. And arches able to read tags have been installed in select sites.[61] But the vision of a technologically secured future, such as that presented to forum attendees in Tlaxcala, remains a distant reality as the REPUVE is blended into existing modes of governance and governmental infrastructures—tolling systems in Sonora and customs houses at the border—in order to make the program work. Rather than unifying the identification of individuals under a single biometric standard, the Personal Identity Card for Minors adds another level to personal identification and more bureaucracy to its governance. And the Geolocation Law fails to bring governmental agencies together at all, casting the responsibility for security over mobile telephony outside the state. Thus, the power of the state is not a static force or entity that maps onto the state's political blueprints, but a stochastic one that emerges through authorities' statecraft with other actors, forces, and modes of governing.

In the end, the true importance of statecraft lies in what it reveals about the continued influence of civic engagement in the formation of the security state. In Sonora, political pressure from ordinary citizens and local politicians translated into the realization of a constitutional right that had previously been ignored. And at the border, that same pressure resulted in the suspension of import fees. Thus, the story of surveillance technologies is one in which ordinary people can remain central to the outcomes that these technologies help produce. In this sense, statecraft can be read as an opportunity for ordinary Mexicans to continue to have a voice in the shape of government. And it is to this point that the concluding chapter turns.

CHAPTER 6

Grasping Surveillance

It's not easy to believe in the government. But we have to
believe in something. We need to come together to make
the government better, to trust it more. I have to take on
my responsibility independent of whether I believe in the
government or not. We have to meet our responsibility. So,
I see this program independently of whether the authorities
do what they're supposed to. We as citizens should fulfill our
obligation. At the end of the day, we have to think of the
future, in our welfare, independent of the difficulties. And
that means acting with values, involving ourselves in social
activities and programs. Without participation, it would be
worse for everyone.

—Zacatecas resident registering with the REPUVE

6.1 THE MORE THINGS CHANGE . . .

Having left office at the end of 2012, Felipe Calderón and his crusade
against insecurity have passed from the public stage in Mexico. But the
problem of insecurity has not. During his campaign and first years in
office, Enrique Peña Nieto sought to shift the public's attention away
from security issues and toward economic and social policy. The hall-
mark of this effort was the Pact for Mexico, an accord signed by the
president and leaders of the three major political parties to put aside
political differences and move the country forward through cooperation in five key areas. These included agreements for (1) "a society of
rights and liberties," which "achieves the inclusion of all social sectors
and reduces the high levels of inequality that exist today between the
people and regions of our country"; (2) "economic growth, employ-
ment, and competitiveness," whereby the "state should generate the
conditions that permit for economic growth that results in the creation

of stable and well-paying jobs"; (3) "security and justice," whose "principal objective . . . will be the recovery of peace and liberty to diminish violence"; (4) "transparency, accountability, and combatting corruption," which recognizes that "transparency and accountability are two tools of democratic states to elevate the confidence of citizens in their government"; and (5) "democratic governability," in which "the political plurality of the country is a undeniable reality derived from a long and incomplete process of democratic transition."[1]

While the pact was criticized as an antidemocratic measure bypassing the authority of the Congress,[2] it did help set a different tone for the new government. And the Peña Nieto administration built upon the pact by passing education reform aimed at increasing assessment of student learning and teacher training; telecommunications reform seeking to break media monopolies; and energy reforms designed to modernize the oil sector by privatizing Mexican Petroleums (PEMEX), the state-owned oil company that is a symbol of national identity dating back to Lázaro Cárdenas's nationalization of the country's oil reserves in 1938.[3]

Reality, however, has not followed the president's script. According to federal crime statistics, homicides have supposedly decreased since Peña Nieto took office. But independent reporting has found the rate consistent with the Calderón era, with over fifty-seven thousand deaths recorded in the first twenty months of the Peña Nieto administration.[4] And if the Pact for Mexico succeeded in capturing the public's attention during this time, the disappearance of forty-three students from the Raúl Isidro Burgos Rural Teachers' College of Ayotzinapa in September 2014 dramatically disrupted the federal government's efforts to manage the public's perception of insecurity. The kidnapping and presumed assassination of the young men who had dedicated themselves to careers in teaching, carried out by the local mayor in conjunction with police forces and a local crime syndicate, rekindled the wrath of a public fed up with the state's complicity in crime. The crimes, together with the inability of state authorities to locate the students' bodies, fueled demonstrations across the country under the banner of "Fue el Estado!" (It was the State!). In response, Peña Nieto did what Felipe Calderón and Vicente Fox had done before him: he announced the creation of a new federal police force—the National Gendarmerie—styled after France's and Chile's militarized national police forces, which would regain territory lost to organized crime through the increased use of cutting-edge technology and intelligence gathering.[5]

Outside Mexico, meanwhile, adoption of surveillance technologies to combat insecurity continues apace. Regionally, the problems of violence and organized crime plaguing Mexico exist in other Latin American countries, and national governments have turned to anonymized mobile device reporting, vehicle control systems, integrated telecommunications networks, video surveillance cameras, and the like in response.[6] In the United States, the killing of innocent people by drone strikes in the Middle East, ongoing revelations about the National Security Agency's massive domestic and international spying operations, and the use of excessive force by local police forces have drawn criticism. This criticism has prompted the federal government to define the use of drones for targeted killings,[7] limit domestic data collection,[8] and reduce the transfer of used military equipment to domestic police forces.[9] But reliance on surveillance technologies against insecurity remains. Globally, national governments use surveillance technologies in many of the same applications described in this book, and authoritarian regimes buy wares from US, Canadian, and European companies to monitor and punish dissenters who are defined as security threats.[10]

With these trends as a backdrop, what lessons does this examination of the Calderón administration's RENAUT, CEDI, and REPUVE programs hold? This concluding chapter attempts to answer this question by reviewing four thematic binaries central to understanding surveillance technologies and the state: visibility/tactility, strength/weakness, determinism/emergence, and fatalism/engagement. These ideas, taken together, underscore that while surveillance technologies might envision a future of tighter governmental control through grabbing hold of the materiality of society, the structure of society that has taken shape over the course of modernity ensures that a space for political action remains, which opens up opportunities for the citizenry to shape the fate of surveillance technologies and governance in the future.

6.2 VISIBILITY AND TACTILITY

Thinking on surveillance tends to privilege sight as a human sense. This is understandable. A fairly recent term dating back to the French Revolution's Reign of Terror, when surveillance committees were formed to monitor suspicious people and political dissidents, "surveillance" derives from the French prefix *sur* (over) and root *veiller* (to watch) and means "to watch over."[11] It was in this sense that Michel Foucault used the word in *Discipline and Punish* (whose French title is

Surveiller et punir), the seminal work that helped popularize the term in the academy.

The emphasis on visibility and sight has endured in our imaginations. Recent scholarship in surveillance studies has shifted this understanding, however, by describing how information technologies such as radio-frequency identification (RFID) tags, biometric cards, mobile devices, personal computers, and the networks that link these devices have transformed surveillance into "dataveillance."[12] The histories of the mobile telephone registry, personal identity card, and automobile registry in Mexico provide detailed case studies of the technical and administrative procedures required to collect data on communications, personal identity, and mobility. And what these cases show is that surveillance technologies operate not only through *visibility* and *watching over people,* but also through *tactility* and *taking hold of and remaining in touch with the materiality of both people and things.* Creating a national identity card based on biometric data requires that the human body be probed and contacted in different ways. Fingers need to be touched and recorded. Irises need to be scanned. These data are then encoded into bar codes and other formats that are stored both in the card and the digital databases of the government. Those databases of the state must then be integrated to eliminate redundancies. Creating a national automobile registry requires that the body of the car be examined, inspected, and touched in order to record three instances of a vehicle identification number inscribed on it. That information is then scanned into government databases and inscribed into RFID tags that are applied directly to vehicles' windshields. The public and private databases related to automobility are then merged to ensure "legal certainty."

This emphasis on touch and adhesion is why it is meaningful to speak of *prohesion* rather than surveillance. If surveillance is understood as "watching over people" for the sake of affecting their behavior, the histories of surveillance technologies in Mexico reveal an operation in which authorities use technological means to manipulate the stickiness or viscosity of the things that energize social life so as to better order society. With these technologies, authorities in Mexico continue an effort dating back to the founding of the nation to manage the materiality of communications, identification, and mobility.

The distinction between visibility and tactility is important for understanding the logic of governmental power today. For the state authorities of the eighteenth and nineteenth centuries studied by Foucault, surveillance and the constant monitoring of people allowed behaviors to be observed, comparisons between individuals to be made, ranks to be assigned, and knowledge to be generated that formed the basis of diverse disciplines or fields of social-scientific expertise. "In short," Foucault noted on surveillance, "it normalizes."[13] Through this operation, surveillance provided the basis for *discipline,* for ordering the chaotic masses of the natural and social worlds into individualized subjects and units. For federal authorities in Mexico who sought to realize the National Registry of Mobile Telephone Users (RENAUT), Citizen Identity Card (CEDI), and Public Registry of Vehicles (REPUVE), *prohesion* enabled registers of the objects and subjects circulating in society to be generated, evidence of their existence to be recorded, a connection to their materiality to be established, and comparisons between those things and officials records to be made. This is not a power interested in individualizing and normalizing the masses, as those individuations have already been made. It is rather a power seeking to match those objects and subjects that circulate in society with the data that exists about them and to localize them or ascertain their presence at a particular time and place. Prohesion, then, allows for the authentication of both people and things. And by this operation, prohesion provides the basis for *security,* for holding onto or preserving the order of subjects and objects in the world as it is.

The distinction between discipline and security has been drawn before, if not in these terms. Foucault already in 1978 described security as a third form of power distinct from sovereign and disciplinary power.[a] What Foucault termed security can be equated to what Gilles Deleuze referred to as "societies of control," where "we no longer find ourselves dealing with the mass/individual pair" present in the disciplinary society—"individuals have become 'dividuals,' and masses, samples, data, markets, or 'banks.'"[14] The dataveillance technologies of the control society are used to "social sort"[15] individuals in countless

---

a. "Baldly," Foucault writes, "we could say that sovereignty is exercised within the borders of a territory, discipline is exercised on the bodies of individuals, and security is exercised over a whole population," where population "will be considered as a set of processes to be managed at the level and on the basis of what is natural in these processes." Put more plainly, security for Foucault is liberal governance, where the state intervenes in social relations so as to create "natural" relations that will provide the conditions for the organic growth of the economy, health, and so forth (Foucault, *Security, Territory, Population,* 11).

social settings: safe/legitimate and dangerous/illegitimate travelers at borders,[16] desirable and undesirable citizens on the streets,[17] automobility and pedestrian mobility at urban intersections,[18] good risks and bad risks for criminal rehabilitation in courts and prisons,[19] and so on. In Mexico, the phone registry, personal identity card, and automobile registry were launched with security as the explicit goal. Authorities wanted to sort between legitimate phones and stolen devices, suspicious and reputable individuals, and dubious and trustworthy motor vehicles.

But if this has been said before, examination of the Mexican government's attempts to implement prohesive technologies raises additional points. Significantly, discipline and security exhibit different concerns on the part of authorities relative to the worlds they look to govern. Discipline entails a missionary logic of transforming and ordering an external world thought to be defined by chaos, disorder, and danger. In the face of the plague, the healthy individual can be created. Out of the unimpressive military recruit, the efficient soldier can be crafted. From the untrained child, the educated student can be molded. From the common criminal, the reformed citizen can be made. Through the artful application of disciplinary techniques—enclosure, partitioning, functional sites, ranks, examinations, time tables—whatever mass of social or natural material can be broken down and remade into individual, productive units. Security, in contrast, carries a custodial logic of preserving that order or advantage that has been won over the world. In the face of terrorist or criminal risk that would disrupt the social order, the terrorist can be sorted out to preserve the status quo. In the face of environmental risk that would threaten the natural conditions necessary to maintain the population, the pollutant can be identified and neutralized to protect the natural order. In the face of disease risks, the infected person can be isolated to maintain the health of the population as a whole. Through the artful application of security techniques—the recording of identities, the tagging of bodies, the monitoring of information, the analysis of statistics—whatever collection of ordered elements can be preserved from risks and threats.

A conservativism is present with security, a fear or anxiety of loss, that is absent with discipline. Discipline is oriented outward and toward the future; it sets out into the world to colonize and conquer. Security is oriented inward and toward the present;[20] it sets up apparatuses to keep the world as it is. In contemporary society, a culture of insecurity reigns, which produces "the insecurity subject" who "is afraid but can effectively sublimate these fears by engaging in preparedness activities."[21]

In Mexico, the context of insecurity breeds a fear that automobiles can easily be stolen, that mobile telephones can be taken and used to extort money, and that family members can be kidnapped. Security measures are intended to provide the confidence that individuals will be able to maintain their hold on these valued items and their place in this valued social order.

More importantly, the distinction between surveillance and prohesion illustrates how discipline and security differ with relation to subjects. At its core, discipline involves subjectification—creating enclosures, partitioning people, and erecting functional sites where constant surveillance provides the means for shaping the human soul and creating the subject. Mexican authorities in the early twentieth century pursued roadway safety by responsibilizing motorists, by requiring them to pass driving tests, mark registration numbers on their vehicles, and carry infraction booklets to enable monitoring by police officers. But security is largely indifferent to human subjectivity. At its core, security involves conservation—creating inventories of things, tagging each one, and keeping them monitored through prohesion to protect the social order that modernity has brought forth. Mexican authorities today pursue automotive security by certifying motor vehicles, inspecting their vehicle identification numbers, and tagging them with RFID chips to automate monitoring by electronic scanners.

In contrast to discipline and surveillance, security through prohesion casts its focus beyond the human subject and its soul to the materiality of things that underlie collective agency in society. To stop the terrorist or criminal, security through prohesion would disable the automobiles, phones, and weapons that enable wrongdoing. Such a strategy matches what has been termed "targeted governance,"[22] where problems such as alcoholism are managed through drug interventions that target specific aspects of the person's biological being rather than more holistic (and complicated) interventions that seek to discipline the self. Prohesion combats crime through the targeted governance of telephones and cars rather than more holistic interventions against the norms and conduct of persons.

A certain distrust of the human subject is detected here—individuals cannot be trusted to preserve the social order themselves. As Benjamin Goold has noted, "The increased use of surveillance technologies might send a particularly negative message to members of the public about how the state views them and the extent to which they can expect the state to trust them."[23] If everyone is a suspect, the simplest way to secure society is to connect the circuits of control directly to the materiality of collective agency.

As the case studies of monitoring programs in Mexico demonstrate, this distrust extends to the state itself. In addition to adhering sentinels to the materiality of collective agencies, prohesion also attempts to integrate the state agencies that have emerged over the course of modernity to govern society. State authorities in charge of telecommunications, tax rolls, automobile licenses and registrations, voter rolls, population rolls, and so forth are made to cohere to one another to improve the state's hold on collective agency. But whereas the "interoperability" and "integration" of monitoring systems[24] are often perceived as an indication of the potency of dataveillance, they here speak to the lack of trust in authorities by authorities.[25] In Mexico, this lack of trust is pronounced. State officials openly say that police officers and other state employees cannot be trusted to carry out the law and protect the social order.[26] The telephone registry, personal identity card, and vehicle registry are ways in which the governance of telecommunications, personal identity, and automobility can be streamlined to increase efficacy. In a sense, then, prohesion evidences a belief that humans, be they the governed or the governors, simply cannot be entrusted with that which security aims to preserve.

As Foucault noted on multiple occasions, the presence of security as a new mode of power does not signify the passing of discipline or sovereign power. They coexist. Nevertheless, the shift to security with prohesion as the means for carrying it out would have serious consequences. Operating by attaching to the substance of our daily lives, prohesion can be particularly invasive. Personal privacy is under assault in various ways under the new surveillance, as the details of our lives get collected by private companies specializing in data management, are traded between public and private entities, or are hacked by digital criminals. Security can also be unjust. The poorest and most vulnerable in society are surveilled the most.[27] As a consequence, the divisions between the haves and have-nots are reinforced, an outcome that aligns with the conservative logic of security to preserve the social order.

In addition to invasions of privacy and the deepening of social inequalities, prohesion reveals a further, more worrisome dimension of security. In its aversion to human subjectivity, prohesion threatens the individual subject. Discipline sought to mold human subjectivity through constant attention to the minute details of people's lives. It represented a culmination of sorts in a "great tradition of the eminence of detail, [in which] all the minutiae of Christian education, of scholastic or military pedagogy, all forms of 'training' found their place easily enough" in the disciplinary society.[28] Security, however, disregards the toilsome, costly, mundane work of keeping watch over people in favor of simply attaching to the materiality of society. As a result, the formation of the subject is no longer a priority. Others have noted an analogous dynamic in speaking of the "data doubles" and "doppelgangers"[29] that dataveillance creates and acts upon in place of physical, autonomous subjects.[30] As Charlotte Epstein has put it, "When the human body is no longer so clearly upheld as the recipient of rights, as the subject of politics, it is not so clear that it is anything more than just a living object, or indeed an animal-to-be-managed."[31] In security, people are reduced from political subjects to physical bodies to be administered.

Beyond this, basic elements of the liberal political order designed to promote subjectivity find themselves under assault in security. In attempting to secure the social order through materiality rather than subjectivity, prohesion alters the individual's grip on the world in subtle but fundamental ways. For one, choice is moderated by mandatory actions that are required in the security society. The REPUVE requires motorists in Mexico to enroll in the automobile registry and adhere RFID tags to their vehicles. The CEDI requires citizens in Mexico to possess personal identification cards. And the RENAUT requires mobile telephone users in Mexico to register their phone numbers with the government. The cost of not doing so is the risk of not being able to access key services central to daily life in contemporary society. Drivers who do not register their vehicles could be restricted from accessing roadways activated by RFID stickers. People without identification cards could be denied social services. And callers who do not register their phones could be threatened with cessation of their cellular service. In the same way, air travelers throughout the world have little choice but to comply with nebulous requirements to publicly disrobe at security checkpoints and even less power to remove their personal communications and data from governmental and private-sector databases.

Second, property rights are slowly chipped away as the state seeks to attach itself to the things of daily life. Drivers in Mexico are mandated to have state-issued RFID devices adhered to their windshields, with little choice as to where the admittedly unsightly sticker is placed. The stickers are present and registered with the state at the point of sale, they cannot be legally removed, and they must be replaced if the windshield is replaced. The windshield ceases to belong to vehicle owners in the way it once did. Consequently, while drivers have never possessed their vehicles entirely (the plate that legally identifies the car belongs to the state and laws commonly proscribe tinted windows and other modifications), the state's placement of RFID stickers colonizes a new portion of the automobile—the windshield—which further limits ownership. Similarly, mobile telephones that are not registered with the state or do not comply with protocol requirements are denied access and operability, thus requiring the purchase of a new device that is already connected to networks of control. Vehicles and telephones still belong to their rightful owners, but in attempting to secure these objects, users are required to surrender aspects of ownership to the state and programs that would protect them.

Third, self-determination is restricted by biometric identification. Electronic identity cards that identify individuals according to their biological material rather than their names result in a diminished space for individuals to define themselves before authorities. This can be seen as an extension of a long trend in Mexican history. Indigenous peoples of Mexico were forced to identify themselves within the naming practices and structure of Hispanic society. But under security, even that diminished capacity to name oneself is removed. With biometric information, one's biology "anchors" identity.[32]

Thus, security by prohesion—by diminishing choice, private property, and self-determination—threatens those fundamental elements of liberal society that ensure subjectivity. And the modern liberal subject is left at risk. Paradoxically, then, if the disciplinary society and visibility carried the goal of subjectifying society, then the tools being used to defend that social order, that subject, and the material things by which it defines itself serve to slowly extinguish the subject.[b]

b. This concern resonates with arguments that critical theorists of a generation ago made concerning technology. The "Megamachine" of modern industrial society, cautioned Lewis Mumford, would eventually "reduce all forms of life and culture to those that can be translated into the current system of scientific abstractions, and transferred on a mass basis to machines and electronic apparatus" (Mumford, "Technics and the Nature of Man," 315). But an important distinction can be made. While the Megamachine and Technique (Ellul, "Technological Order") reduced the subject to one dimension (Marcuse, *One-Dimensional Man*), they still required a substantial investment of human action and oversight in order to cultivate that dimension. With security, the subject is bypassed altogether and the conditions under which she or he would develop, even along a single trajectory of technical specialization and market consumption, are restricted.

6.3 STRENGTH AND WEAKNESS

If security through prohesion offers a troubling vision of the power at work in security surveillance technologies, solace can be found in the fact that this power encounters such difficulty in taking root. Of the three programs examined in this book, one was abolished by the Mexican Senate because of its failings, one is stuck in limbo awaiting action from the Peña Nieto administration, and one is operating in a weakened form that fails to fulfill the vision of automobile security intended in its design. In this sense, weakness is a central aspect of security and prohesion in Mexico.

Failure is a topic that surveillance scholars have treated in the past. The surveillant state has been referred to as the Big Bungler rather than Big Brother, an authority "driven mad by too much power and too much speed."[33] Errors are common in the data that public and private entities gather about us, which "can lead to death in hospitals, stolen elections, and wrongful arrests."[34] The substance of life itself can throw security technologies off. Facial-recognition technologies are doomed to fail "since identity is inherently a hybrid and unstable construct—at the very least, individuals age, take different jobs, acquire and lose credentials, marry and divorce, etc.—it can never be completely and absolutely stabilized."[35] And multiple standards for the recording and storage of information can spoil government attempts to implement a national identity card.[36,c]

But if failure has been recognized in the literature, perhaps it has not received the emphasis it deserves. Within society, we feel either trepidation or relief, depending on our political affiliation, when government designs for surveillance are announced or leaked to the public. And this reveals the confidence we have in these plans. Militarized drones unsettle us because they illustrate how the conduct of warfare and killing is escaping human control and becoming automated. The unimaginably vast snooping activities of the US National Security Agency (NSA) revealed by Edward Snowden, Glenn Greenwald, and Laura Poitras concern the critical minded of us because they imply that the minutiae of our daily phone and electronic communications are open to inspection. The adoption of national identity cards disturbs us because it

---

c. These failures have not, however, turned governments off of surveillance technologies. As Clive Norris has noted, "nothing succeeds like failure" when it comes to using technology in the pursuit of security (Norris, "Success of Failure").

signifies the erection of new walls and boundaries that will break our contact with the Other and endanger our free society. In short, our fears about the negative consequences that accompany surveillance technologies rest on the assumption that these technologies have the strength they claim to have. And in the face of this power, as the move to adopt the legal concept of the "right to be forgotten" in the European Union demonstrates, all we as concerned individuals and groups can do is ask that this power be fallible, that it forget.

It is beyond debate that technologies in contemporary society carry a capacity for tracking and oversight unlike anything that has come before. Militarized drones are certainly unleveling the playing field for the conduct of war. NSA surveillance over personal communication, Big Data or otherwise, is an affront to the notion of a free society. Biometric identity cards are a technological step in the direction of increased control over personal identification. And these technologies do sometimes succeed in assassinating suspected terrorists at a distance, scooping up critical pieces of information to stop a crime, or achieving access control. But the continued insecurity of our world speaks to a fundamental weakness or fallibility of security systems.

Perhaps the most telling example in this regard is the Boston Marathon bombings, where the brothers Tamerlan and Dzhokhar Tsarnaev exploded two homemade bombs at the finish line of the foot race on April 15, 2013, killing three and injuring scores of others. Lost in the tragedy of the event and the drama of the subsequent manhunt is the fact that the multiple surveillance programs and various layers of surveillance technologies instituted since the September 11, 2001, terrorist attacks failed to identify the two brothers as threats. This despite the fact that they were born in the conflict-torn Caucasus region of the Soviet Union, self-identified as Chechen, had previous encounters with the police for violent behavior, and learned bomb making from an online magazine published by al-Qaida. What is more, following the attacks, Senators Saxby Chambliss and Richard Burr reported that Russian intelligence officials had warned both the FBI and CIA about the brothers, including recordings of Tamerlan discussing attacks with his mother over the phone.[37] So, then, not only did the "surveillant assemblage" fail to capture these terrorists, but, to invoke a Marxist argument, it might be argued that these technologies have "deskilled" traditional intelligence work to the point where information provided by another country's intelligence service was not acted upon in the manner one might expect.

Similarly, the brothers Cherif and Said Kouachi, who killed twelve and injured eleven during an attack on the offices of the satirical magazine *Charlie Hebdo* in Paris in January 2015, had been under surveillance by French authorities; Cherif had even been arrested and tried on terror charges in 2005 as he was heading to Iraq to fight US forces. Thus, authorities in France, who possess some of most sweeping powers to surveil the public and regularly deport alleged extremists without the procedural protections of the US legal system, were unable to prevent this attack.[38] Zarrar Shah, the technology chief of Lashkar-e-Taiba, the Pakistani terror group that carried out a series of coordinated attacks in Mumbai over the course of three days in November 2008 that left 164 dead and 308 injured, used Google Earth to plot the attacks and was being monitored by British, Indian, and US authorities. Yet, the surveillant assemblage proved too weak to stop these attacks.[39] Ismaaiyl Brinsley, the gunman who ambushed two New York City police officers in December 2014, had earlier in the day shot his girlfriend in Baltimore. Baltimore police, using pinging technology to locate Brinsley's mobile phone, notified New York City police that he was in Brooklyn and was posting messages on his girlfriend's Twitter account saying that he would kill two New York City officers.[40] But this, too, failed to stop the attack. And the events that bookend the birth of the massive US homeland security state—both the September 11, 2001, terrorist attacks and Edward Snowden's whistleblowing about NSA domestic spying—speak to the failure of surveillance. Multiple agencies had information about the September 11 attackers, but this information was not acted upon. And Snowden's revelations demonstrate the permeability of a surveillant assemblage that relies on private firms to provide public security.

Mexico, meanwhile, was rocked in 2014 by the disappearance of the forty three Rural Teachers' College students in Iguala, Guerrero. A federal investigation implicated the mayor of Iguala and local police. The investigation found that the police had apprehended the students and turned them over to a local crime syndicate, Guerreros Unidos (United Warriors), which then presumably murdered them. Incredibly, despite the immense investment of technology and resources in the fight against crime, the federal government was unable to locate all but one of the students' bodies.

The legality and desirability of intrusive surveillance technologies in our lives will continue to be debated. But if these technologies are already operating, they might be expected to work at least at modest levels. As these examples show, however, security surveillance and prohesion not only sometimes fail but are fundamentally weak forms of protection.

The Registry of Mobile Telephone Users, Citizen Identity Card, and Public Registry of Vehicles pursued by the Calderón administration provide insight into the forces that account for the weakness of the weapons of the security state. First, apart from the technologies themselves, the turn to surveillance technologies speaks to a distinct weakness of government. In Mexico, the state simply cannot govern the way it once did. The elevated levels of ordinary crime, the immense numbers of homicides, the underreported number of femicides, the common kidnappings, and the arms and drugs trafficking all illustrate the inability of the state at both the federal and state levels to provide security.

Chapter 2 discussed the reasons for the weakening of the state and strengthening of criminal elements in Mexico. The dictatorial, single-party rule of the Institutional Revolutionary Party (PRI), whatever its shortcomings as a democratic form of government, provided a centralization of political power that proved able to manage drug trafficking and the violence that can accompany it. Democratization has brought about free, competitive elections at different levels of government and increased civilian control over the political process. But this progress has changed political dynamics in the country, decentralizing power and weakening the clientelist relationships that historically corralled drug violence.[41] At the same time, the death of Amado Carrillo Fuentes, leader of the Juárez cartel, the original *jefe de los jefes* (boss of the bosses), precipitated the current and ongoing wave of violence because it created a power vacuum that various regional cartels and criminal organizations sought to fill. The lack of a monopoly over criminal activities in Mexico by either the state or crime bosses has resulted in a rise of formerly unauthorized forms of violence, such as kidnappings, extortions, and street robberies.[42]

These transformations in Mexico's political landscape were accompanied by changes in the country's economy. The shift from a statist, protectionist economy controlled by the PRI to a neoliberal political economy governed by free-market policies has expanded the gross domestic product and enriched Mexico's upper and upper-middle classes as well as regions along the northern border.[43] But this wealth has not been shared equally; the poverty rate (as measured by income required for basic living expenses) has remained stuck at 50 percent of the population,[44] indicating increasing income inequality. Crime, then, has become one way for people living at the margins of society to pursue economic gain.

Thrown into this social mix is the transformation of Mexican cultural life through exposure to global media, which simultaneously weakens certain forms of traditional national identity while strengthening others—*pulquerías* and *siestas* gradually disappear as tastes and times change in concert with global norms, while *narcocorridos* that glamorize and romanticize the fatalist pursuit of drug wealth rise in popularity as a distinctly Mexican form of cultural expression. Together with an active feminist movement[45] pushing for reproductive rights and other protections, as well as other forms of global consciousness, these changes weaken the legitimacy of traditional authorities and ways of doing things. Thus, over the past decades, the Mexican state has contracted in accordance with the precepts of neoliberal governance, which has reduced its ability to govern, while the society it oversees has continued to expand, evolve, and transform as it absorbs new technologies and means of expression and it experiments with new freedoms presented by democratic governance. With less ability to govern, and an unreliable police force with which such governance could not be entrusted, the Mexican government turned to surveillance technologies as a way to reform itself to govern in a global world.

Second, the national government's failure to fully implement surveillance technologies has shown that it is prone to weakness. Resistance has been central in this regard. Resistance meets authorities' efforts to create the security state at various points. Mobile phone users suspicious of the federal government's registry refused to register their lines honestly. And the poor design of the registry left it unclear how users' phone lines could be verified and who would even have the responsibility for doing so. Drivers unaware or uninterested in the federal government's automobile registry in the states where it was being offered failed to register their vehicles. And many states refused to participate in the program altogether, their reluctance motivated by politics and a fear of wasting precious security resources on a flailing federal program. The Citizen Identity Card failed to launch due to opposition from the government office responsible for issuing a rival identity card.

Resistance is an established topic within surveillance studies. John Gilliom, for instance, in his examination of an electronic payments system that monitors public assistance in Ohio, demonstrates how poor women's defiance of welfare rules constituted an everyday form of resistance that opposed the power of the state as "overseers of the poor."[46]

And Gary T. Marx has provided an authoritative accounting of the myriad ways in which people resist everyday forms of monitoring, such as drug testing in the workplace, a list that includes "refusal" (to take a test), "discovery" (of the date of a random test), "avoidance" (not going to work on testing day), "switching" (a clean drug sample for a tainted one), "distorting" (consuming substances to neutralize the drug test), "masking" (one's identity to testers), and "countersurveillance" (testing on oneself to ensure success)." Marx observes that such strategies "should serve as humbling reminder of need for skepticism in the face of unreflective paranoia and oversold technical surveillance fixes introduced into heterogeneous social contexts."[47]

Supporting Marx's conclusion, the histories of security surveillance in Mexico encourage a broader definition of resistance—any force, whether human or not, that has the effect of obstructing the intended plans and intentions or established relational patterns of authorities (see chapter 4)—to take fuller account of the variety of difficulties inherent in establishing new modes of oversight and governance in society. It is not only that people, whether private citizens, CEOs, or elected officials, oppose these tactics and authorities. But time, space, and the technologies themselves intervene as well. Given these diverse forces, prohesion fails to acquire the power that it was designed to possess.

Implicit in this definition of resistance and central to understanding the weakness of surveillance technologies are the concepts of "distributed agency"[48] and "assemblages"[49] introduced earlier in this work. A car is not simply a car, a phone is not simply a phone, and a person is not simply a person. They are rather elements situated in a larger network of associations between people, organizations, things, and ideas that enliven them. This is "vibrant matter."[50] And having authorities take hold of those things—phones, people, automobiles—in turn means engaging with the range of associations that give them agency. The RENAUT, CEDI, and REPUVE largely failed to take hold of mobile telephony, personal identification, and automobility in Mexico because these collective agencies are distributed across a wide network of users, providers, regulatory agencies, and material things that enables their activity. To get a grasp on mobile telephony, it is not enough to simply have users register their numbers with the appropriate authority. Mobile service providers, the governmental agencies regulating telecommunications, the designers of phones, the placement of cellular towers, and so forth must be integrated into the program as well. To take control of the wheel of automobility, the government must ensure not only that car companies provide records of sales to the government's database and adhere RFID stickers to windshields, but also that state governments and customs officials do the same with vehicles circulating in the country or crossing national borders.

Daniel Neyland, in an innovative examination of governmental efforts to control "everyday objects of terror"—letter bombs, sharp objects and liquids on airplanes, and so on—makes a similar point about the inherent difficulty of securitizing things. "The example of objects in airports," Neyland notes, "suggests that successive actions to build networks of governance around categories of objects (such as liquid containers and sharps), connecting various people (airport managers, passengers, security and check-in staff) and things (boards, plasma screen TVs, leaflets) in order to reorient actions around the object in focus and establish its new ontological status as a matter of concern are messy in practice." Quite simply, he concludes, "it seems that ontologies are stubborn and routinized."[51]

What is most interesting about the ontological stubbornness of things is the manner in which older structures of governance get in the way of newer ones. The principal opposition to the Citizen Identity Card came from the Federal Electoral Institute (IFE). The main challenge to the Public Registry of Vehicles was the opposition or lack of participation of the states. Both the IFE and the federated states of Mexico are bodies that govern in Mexico. Historically, they emerged as authorities worked to solve particular problems of governance that faced the nation. The IFE was created to provide legitimacy to a fledgling democratic electoral system that did not have the trust of the public following the dubious presidential elections of 1988. The states came into existence as a means for governing Mexico's outer territories of that could not be effectively ruled by centralized authorities, giving birth to "the negotiated state."[52] These are state forms that were "co-produced"[53] over time in conjunction with those things and phenomena they were designed to govern. However, the security state encounters them as obstacles that prevent the implementation of prohesion. These thoughts cast in sharper contrast the weakness of weapons whose strength authorities are always assuring us of.

6.4 DETERMINISM AND EMERGENCE

But to say that surveillance technologies are fundamentally weak is not to say that the state in Mexico lacks power. Through these programs, federal authorities can require *sujetos obligados* (obligated subjects) like automobile manufacturers and *entidades federativas* (federated entities) to deliver data about the production, sales, and registration of vehicles to the REPUVE database; local, state, and federal law enforcement use the database to search for and identify stolen vehicles; states such as Sonora are able to employ RFID technology as a tolling solution or the basis for tax collection; and this progress provides the federal government a basis for further extending this surveillant assemblage into other states and state agencies in the future. The federal government has also been able to distribute four million personal identity cards to schoolchildren in several states throughout Mexico. Even with the failed mobile telephone registry, the state was able to register nearly eighty-three million mobile phone numbers, or 90 percent of all numbers in Mexico; and when the registry was ultimately terminated, the federal government succeeded in quickly transferring responsibility for monitoring telecommunications to service providers.

These outcomes and this arrangement of power, however, are not what the state had planned. This is not the secure future that prohesion as a novel form of governmentality promised. It is rather the unexpected result of authorities negotiating with the people, organizations, rules and laws, things, and concepts that had resisted the programs. This arrangement of power is, as noted in the last chapter, the product of statecraft.

The improvisational character of social life has been highlighted by several influential works in the social sciences. The best-known version of this idea is "bricolage," which Claude Lévi-Strauss used to denote tinkering or "someone who works with his hands and uses devious means compared to those of a craftsman"[54] in order to distinguish premodern forms of knowledge from their modern, scientific counterparts.[d] In a

---

d. "The bricoleur is adept at performing a large number of diverse tasks," Lévi-Strauss claims, "but, unlike the engineer, he does not subordinate each of them to the availability of raw materials and tools conceived and procured for the purpose of the project. His universe of instruments is closed and the rules of his game are always to make do with 'whatever is at hand'" (Lévi-Strauss, *Savage Mind,* 17). This notion of making do with whatever is at hand has been adapted to a variety of works in the social sciences, perhaps most apropos to the topics discussed here by Claudio Ciborra, an organizational theorist, who in describing the successes and failures of strategic information systems within organizations, comments that "in order to achieve a new SIS (strategic information system) design the issue is neither to try to generate the most creative application idea, nor to realize the design through a careful planning and implementation method. The real issue is being able to overcome those cognitive and institutional barriers that prevent users and designers [from] seeing, appreciating, and utilizing all those potential applications already surrounding the members of an organization" (Ciborra, *Labyrinths of Information,* 44).

similar vein, Andrew Pickering describes scientific and engineering work as "a mangle of practice," a "practical, goal-oriented and goal-revising dialectic of resistance and accommodation" by which scientific knowledge and technological artifacts emerge in time.[55] And most closely related to the current book, James Scott's research on the state argues that state-initiated social-engineering programs, like the collectivization of Soviet farms or the construction of high-modernist cities like Brasilia, are doomed to fail and that human societies would be better served by governance based on "metis," that is, "folk wisdom" or "knowledge that can only come from practical experience."[56]

Recognizing the presence of tinkering and improvisation in the deployment of surveillance technologies has important consequences for understanding the power of the state. Most importantly, it identifies a skill-based, human component of state formation that cannot be reduced to larger structural forces, be they the authority of rulers, the composition of state power, the accumulation of capital, the culture of a society, or the design of technologies. Such forces clearly mattered in the outcomes of the RENAUT, CEDI, and REPUVE. But the successes and failures these programs experienced had as much to do with the skill of state officials and administrators, like Samuel Gallo, in recognizing an opportunity to connect, for example, the REPUVE to an existing state program and negotiate with those authorities to "make things stick."

And to develop the point further, there is nothing—not the skill of the state practitioner, the authority of the lawmaker, the design of the program, the beliefs of the population, the wealth of the company, or anything else—that can guarantee that a particular modification will actually take. In the case of the REPUVE, some improvisations worked. In the case of the RENAUT, most did not, which left monitoring of mobile telephony in Mexico outside the organizational structure of the federal government. The outcomes of the state's adoption of surveillance technologies to fight insecurity are thus decided in good measure through trial and error.

Over the past two decades, there has been increasing acceptance of the idea that social phenomena do not have singular causes but are "co-produced" through the interaction of various elements. The social order is, in other words, emergent. The concept of "emergence," which is central to science and technology studies and "assemblage thinking," offers a needed exit out of the disabling "structure versus agency" debate in the social sciences.[57] Applied to politics, the concept of emergence avoids having to explain the formation of the state as resulting directly from either the plans of great statesmen or the structure of capital, coercion, or culture.[58] As the second chapter illustrated, central dimensions of the Mexican state took shape over time through authorities' evolving efforts to maintain control over communication, identification, and mobility in society. And as the last chapter recounted, even when plans for reforming the state are known in advance, the shape that reform ultimately takes can only be settled in practice.

These ideas are relevant to surveillance studies. Regularly, works on surveillance give the impression that these technologies are transforming the world in line with their technical design. Security as a mode of governance based on the social sort has arisen because electronic identity cards allow biometric data to be stored simultaneously in the cards and government databases. Security is marked by a diminution of democracy because private corporations are intimately involved in the planning, development, and deployment of surveillance systems, and these companies are not accountable to the public as elected officials are. Personal privacy has already passed into history in the surveillance society, because the bits of information that we are constantly generating through our electronic communications, online searches, plastic-card purchases, and so on are scooped up by public agencies and private-sector actors that use the data without our consent. Statements such as these, simplified perhaps but not uncommon, reveal a determinist mode of thinking where direct lines are drawn between particular social phenomena and surveillance technologies, or where the social consequences of surveillance technologies are predicted in advance. This thinking is not technological determinism. It is technology, in conjunction with multinational corporations or secretive state security agencies, that determines outcomes.

It was this tendency toward determinism that prompted thinking about society in terms of emergence in the first place.[e] And remaining

---

e. Before "emergence," explanations for the creation of scientific knowledge, technological objects, and their impact on the social world were told in the language of the sociology of scientific knowledge or the social construction of technology. These social constructivist perspectives viewed facts, such as those resulting Robert Boyle's pneumatic experiments (Shapin, "Pump and Circumstance"), and artifacts (Pinch and Bijker, "Social Construction of Facts and Artefacts"), such as the design of bicycles, as the result of cultural forces (the interests of scientists, the creation of dissemination outlets with which to publicize research and widen the witnessing of science, the replication of experiments before influential public figures who could lend increased legitimacy to science, the formation of a particular vocabulary for describing science and demarcating it from other fields of engagement with the natural world believed less rigorous, social mores dictating the propriety of dress for men and women, and so on).

sensitive to emergence is vital, since it can reveal processes of social change and state reformation that surveillance technologies may be creating. With this in mind, a few points on the emergent nature of surveillance technologies are in order.

First, we should expect the unexpected. Surveillance technologies might sometimes function according to design. But they should be expected to morph as the practices of statecraft fit them into particular settings. The REPUVE and the CEDI took root in Mexico, but they did so in forms and with functions distinct from those planned by authorities.

Second, the relevance of things is relative. Certain elements of social arrangements that were once unimportant or nonexistent can become central to the governance of society, while others that were once central can become inconsequential. Programs such as the RENAUT, CEDI, and REPUVE are intended to insert new elements—computer software, biometric identity cards, RFID tags—into existing distributions of collective agency to increase the government's hold over communications, personal identification, and mobility. But statecraft can involve unexpectedly giving new purpose to old elements. State planners used the toll plazas already constructed in Sonora to their advantage in order to install RFID readers to serve the REPUVE program, just as they used public schools throughout the country to register schoolchildren for the CEDI. Statecraft can also involve getting rid of old elements that were once central to the social order. Old laminated cards that people once used for tolls in Sonora are slowly passing out of use. And old elements that were never part of an assemblage to begin with, such as the constitutional right to free transit, which was not being respected in Sonora, can gain new life through the alignment of forces that statecraft and surveillance technologies bring about.

Third, problems can sometimes become solutions. It is interesting to consider how the shape of a particular assemblage can have consequences for its governability. All of the surveillance programs described in this work failed to meet their designs. In the case of the CEDI and REPUVE, the main point of resistance that dogged the programs came from the state itself, from the extant political structure for governing personal identity and automobility in Mexico. The RENAUT, however, encountered no such opposition. A structure of state agencies never coalesced around the mobile phone—a more recent technology that appeared when neoliberal political economy had already made regulation a mostly private affair—as it had around personal identity or the automobile or the land-line phone. Counterintuitively, however, the very political structure that inhibited the implementation of the REPUVE could, because of its permanence, later be recrafted by program administrators to make the program stick. The RENAUT, by contrast, having no existing state structure for program administrators to graft onto, was simply terminated, the responsibility for governance turned over to those in possession of the necessary infrastructure: private service providers.

Finally, as emergent phenomena, security surveillance technologies will take different meanings based on the context into which they are fit. In Sonora, the REPUVE is valued nearly universally as a means for establishing and respecting the right to free transit that was fought for and established in the Mexican Revolution. In Zacatecas, the REPUVE is understood and approached more cautiously as another government program promising security. At border crossings, meanwhile, the REPUVE is viewed negatively as another scheme to squeeze tax revenue out of individuals who import their vehicles from abroad. In sum, what surveillance technologies do and what they mean emerge in time and practice. This is how the power of surveillance technologies forms.

6.5 FATALISM AND ENGAGEMENT

Emergence has surprising political consequences. Thinking about surveillance is often tinged with a dystopian outlook that minimizes the potential of individual and collective action to influence a surveillant assemblage composed of national governments, transnational corporations, and advanced technologies.[59] This skepticism is matched by popular reactions to controversies such as the NSA spying programs, reactions that vary from support (belief that surveillance technologies keep society safe), to indifference (belief that people should have nothing to hide), to impotence (belief that surveillance technologies are invasive but nothing can be done about it).

But the emergent nature of surveillance technologies means that individuals, despite the design of prohesion as a mode of governance that would control society by bypassing people altogether, still influence government in meaningful ways. The lowly bureaucrat plays a key role in tailoring surveillance technologies to fit existing assemblages of collective agency. And ordinary citizens, through organized efforts to resist a phone registry, parental expressions of uneasiness about the collection of schoolchildren's biometric data, or mere gossiping about state surveillance, help determine whether and how these efforts stick.

If ordinary people remain central to the outcomes of surveillance technologies in society, what are we to do? Which types of actions might influence the presence of surveillance technologies in our lives? How might "participatory democracy [be] enacted through work in and on material objects" such as surveillance technologies?[60]

A sensible place to begin answering these questions is with the efforts activists are already making to engage the surveillant assemblage. Here, it is appropriate to mention the whistleblowers in the employ of the national security state—Chelsea Manning and Edward Snowden—who brought attention to the operation and scale of state security surveillance by releasing classified information about their work. The actions of these individuals, undertaken with the assumption that their lives would be destroyed, were brave and daring. And they resulted in public awareness about the abuses of the US national security state, an essential first step to broader action.[f] Increasing awareness about the workings of surveillance in the world today is the goal of a wider network of activists as well, including the more academically minded Surveillance Studies Centre housed at Queen's University in Canada and civil liberties organizations such as the Electronic Frontier Foundation and the Electronic Privacy Information Center. These groups have organized to pass key legislation or support litigation establishing individual rights against state surveillance. Representative of this collective labor is the "right to be forgotten" established by the European Court of Justice. The court's ruling in *Google Spain SL, Google Inc. v. Agencia Española de Protección de Datos, Mario Costeja González* provides all individuals in Europe the right to prohibit Google and other search engines from linking to items that are "inadequate, irrelevant or no longer relevant, or excessive in relation to the purposes for which they were processed and in the light of the time that has elapsed."[61]

Efforts such as these concern the encroachment of surveillance on fundamental civil liberties. Generally, the surveillance in question is undertaken in the name of national security or by companies involved in information commerce. Such efforts, then, resemble the organized

f. Indeed, in the wake of the Snowden disclosures, the US Congress decided to phase out the NSA's bulk collection of phone records, and allies of the United States subject to its surveillance have drafted resolutions in the United Nations calling for a cessation of such surveillance.

resistance to state surveillance described in this book, such as the digital mobilization of phone users in Mexico against the RENAUT and the subsequent campaigns against Peña Nieto's telecommunications reform, which activists saw as a threat to net neutrality. Taken together, individuals in these instances can be seen working to ensure *freedom*—to preserve a free space in society unfettered by surveillance technologies, which is a condition for democracy.

These efforts, though, assume that surveillance is unsuitable to any civic purpose. This might seem like a trivial qualification, since the massive sweep of information that takes place under the NSA's domestic surveillance program so clearly violates our sense of basic decency and liberty. But there are many examples in which activists have worked to extend the surveillant power of the state to areas of social life often kept in the dark. A clear example is gender violence, such as intimate partner abuse and sexual assaults, where offenders are enabled by the deference the state has historically paid to family privacy and by the stigma of being a victim of such crimes. While legal measures have been passed to protect women from physical and sexual abuse, the power of such laws often proves ineffective against assailants unafraid of criminal sanction. In response, antiviolence advocates across the United States, for instance, have campaigned for legislation that would establish monitoring programs featuring GPS technology to track abusers who repeatedly violate restraining orders and would alert victims when they are nearby.[62] Using surveillance technology to confront gender violence is relevant to Mexico too, where femicides are a prominent form of crime. To combat them, activists have advocated for the use of information technology and mobile devices to publicize the problem and give potential victims the ability to access help.[63]

As these examples illustrate, the situations where activists might campaign for more state surveillance often involve crime rather than national security or data commercialization. In these instances, people look to extend the surveillant power of the state to provide the protection of the law to individuals who are not receiving it. But like the examples of national security and data commercialization, it is assumed that a rule of law exists in society and that authorities have an interest in extending surveillance.[g]

g. It should be noted that in Mexico, women's advocates have often accused the government of apathy toward victims of femicides.

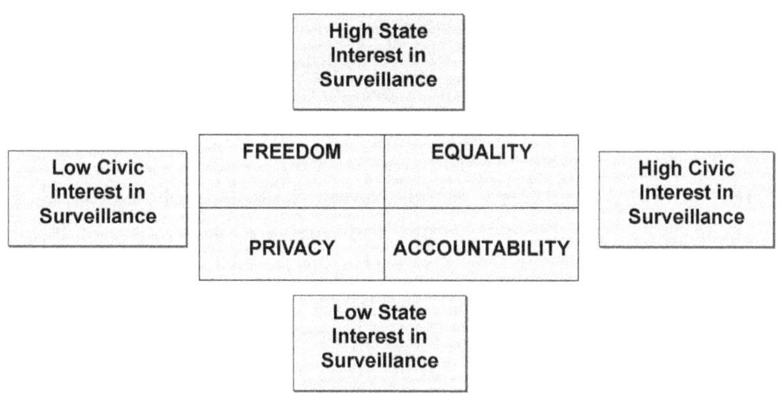

FIGURE 26. Values at stake in surveillance politics.

These considerations help mark out a pair of axes—civic interest in surveillance and state interest in surveillance—against which a politics of surveillance can be measured. Where civic interest in surveillance is low but state interest high, as in the cases of national security and data commerce, activism can be thought to concern *freedom*. Where civic interest in surveillance is high and state interest is too, as in the case of gender violence, activism can be thought to concern *equality*. Those working to end gender violence are interested in ensuring women equal protection before the law (fig. 26).

Campaigns centered around equality reflect what David Lyon has referred to as the "care" dimension of surveillance technologies, at work in hospitals and schools, that accompanies the more discussed "control" dimension. Such campaigns also embody his call for surveillance governed "by an ontology of peace rather than of violence" and "an ethic of care rather than control."[64] They also relate to the "conviviality" of technology that Torin Monahan has called for, describing technologies that "not only afford but also invite modification on the part of users, support diverse modes of expression, and enable power equalization among people."[65]

In contrast to the scenarios involving national security and crime, where state interest in surveillance is a constant, there are others where it is not. In New York City, for instance, public outcry over the conduct of its police force, including the disproportionate use of stop-and-frisk tactics on poor and racial and ethnic minorities, pushed Mayor Bill de Blasio and Police Commissioner William Bratton to implement a pilot program in which police officers wear body cameras to monitor their interactions with the public.[66] While unpopular with the officers, who contend that the cameras will deter people from wanting to talk to them and violate their privacy,[67] police use of such body cameras is expanding in the United States. At the national security level, the US War on Terror has been conducted in a shadowy realm—involving extralegal tactics such as extraordinary rendition, black sites, and secret intelligence court rulings—that activists seek to bring to light.

In these instances, authorities engage in violence—police use of excessive or illegitimate force, the CIA abduction of terror suspects—that they wish to keep from public view. Against these machinations of power, activists use surveillance technologies—body cameras, flight records, maps—to document the illicit actions of the state. In contrast to subjects concerned with *freedom,* who use the rule of law to oppose the state's expansion of surveillance, and subjects concerned with *equality,* who use the rule of law to support the state's expansion of surveillance, individuals here find themselves without a true rule of law. In these settings, they use surveillance technologies to foster *accountability* and legality.

This politically progressive use of surveillance technologies has been pursued by activists in Mexico to document and publicize the assassination of journalists. The map and accompanying database assembled through the Mi México Transparente (My Transparent Mexico) project provides a register of the number and type of attacks suffered by journalists.[68] This register functions as an ongoing surveillant document that announces the threat faced by journalists to members of the state and criminal community who might prefer to silence reporting.

Another innovative use of surveillance technology involved the Yo Soy 132 (I Am 132) movement that captured international attention in 2012 during Enrique Peña Nieto's presidential campaign. In May 2012, the then PRI candidate presented his political platform at the prestigious Ibero-American University, in the prosperous Santa Fe area of Mexico City. During the question and answer session, Peña Nieto angered students when he aggressively defended his actions as governor of the state of Mexico in the 2006 Atenco case, in which hundreds of state police were sent to break up a protest against the planned construction of a new airport. During the police action, two hundred activists were arrested, two were killed, and twenty-six women were sexually assaulted.[69] Following the candidate's response, students broke out with chants of "Assassin!" and "Get out!"

Media coverage of the event downplayed the protest by attributing it to elements outside the university rather than Ibero students, members of one of the more respected institutions in Mexico. Responding to what they saw as the media's attempt to appease the popular candidate's political camp, 131 Ibero students produced a YouTube video showing them with their identity cards as a way of documenting their status as Ibero students and their opposition to Peña Nieto. The video went viral. And supporters of the students responded on Twitter by announcing "Yo Soy 132," or "I am 132," adding themselves to the list of young people against the candidate. Thus, against a media and political establishment that dismissed dissenting voices as disreputable malcontents not worthy of society's respect, the Ibero students and their supporters used the tools of surveillance to announce their presence and opposition to authorities.

Finally, in addition to activists who oppose the state's support of surveillance in pursuit of *freedom*, activists who endorse the state's support of surveillance to fight for *equality*, and activists who support surveillance against the state for *accountability*, it stands to reason that there are contexts in which neither the public nor the state have an interest in surveillance, or at least an interest that would support democratic ideals. This raises what can be called a true sphere of *personal privacy*, where the details of the nonpublic lives of both governors and governed would be respected and not subject to surveillance. The sexual liaisons of public officials (US president Bill Clinton or French president François Hollande come to mind, assuming no crimes were involved) or other details of public leaders' personal lives could be imagined as of no significance for the welfare of the country. And the same assumptions could be made of the intimate personal details of citizens' lives. The fact that there is knowledge about public officials' personal lives or that the state surveils personal aspects of citizen's lives indicates a certain perversion of democratic ideals that has come to masquerade as political controversy.

Nowhere is this more apparent than in the political battles over reproductive rights. The steady push to criminalize abortion in those countries where it is protected under law functions as an effort to increase control over the private lives of women, serving in turn to diminish their capacity to be full subjects in society. And surveillance plays a central role in this contest. The US state of Indiana, for instance, recently considered, although ultimately did not pass, a measure that would have required doctors to partner with and publicize the names of other medical professionals—"backup doctors"—who might treat any complications or emergencies related to an abortion in a nearby hospital. Through such legislation, antiabortion activists sought to publicize the names of doctors who perform abortions, which would presumably expose them to intimidation.[70]

Surveillance over people's personal lives works to the detriment of democratic governance. In these contexts, then, efforts to protect women's right to control their own bodies or to establish that right where it does not exist count as political actions in support of subjecthood. In this regard, the movement to decriminalize abortions in Mexico can be understood as not only an extension of women's rights but also the creation of a social and legal notion of personal privacy that is critical to democracy.

This description of the differing relationships between subjects and surveillance in democratic society is surely too neat. The categories overlap in practice. Many citizens express no concern that surveillance in the name of national security infringes on basic liberties and freedoms. Others would be opposed to the expansion of surveillance in the name of crime fighting, even to combat gender violence, since it would invariably encroach on a sphere of life thought private. Many people consider government secrecy in policing, intelligence, and warfare critical to security. And others believe that freedom of speech provides the legal justification for peering into the private details of people's lives. Quite simply, not all people are the same, nor are all governments the same when it comes to surveillance.[71] But the purpose of this thought exercise is not to close the door to thinking about surveillance technologies, but to open it in order to think about them differently in the hope that they might effect a wider change in how we interact with authorities.

With this in mind, we might return to El Bunker and consider again the architectures of authority found around Mexico City's Chapultepec Park. The subterranean Federal Police Intelligence Center serves as an apt symbol for contemporary approaches to security governance. It operates out of view of ordinary citizens while attempting to remain in contact with them through its array of advanced surveillance technologies. And its technical struggles prove equally emblematic of the failings of this strategy. Historical data are unmanageable, interagency communications are unreliable, state agencies are reluctant to share data, and manual processes of information management at the local level slow data processing and accuracy. It is doubtful that constructing more bunkers will prove decisive in Mexico's War on Crime. If building edifices like Chapultepec Castle above the people bore little fruit in terms of achieving a better society, it should not be surprising that constructing fortresses like El Bunker below them should prove disappointing as well. Only by building structures that require those in positions of power to see eye to eye with those in whose name they govern can a more just and secure future be brought into view.

Notes

CHAPTER 1. SURVEILLANCE TECHNOLOGIES AND STATES OF SECURITY

1. De la Luz González, "'Cerebro' tecnológico enfrentará criminales."

2. Christie Digital, *Federal Police Intelligence Center of the Public Security Secretariat.*

3. "Para combatir al crimen, inaugura presidente el Centro de Inteligencia de la Policía Federal."

4. Reséndiz and Morales, "EPN."

5. Sarakki Associates Incorporated, *Mexico's National Command and Control Center Challenges and Successes.*

6. *Terra,* "Policía Federal muestra interior del Centro de Inteligencia."

7. *La Jornada,* "Construirá el gobierno un búnker secreto para labores de inteligencia antinarcóticos"; Guzmán Roque, "El 'Bunker' más grande de AL es inaugurado por GDF."

8. Lyon, *Surveillance Society,* 2.

9. Marx, "Surveillance and Society," 1–2.

10. Whitaker, "A Faustian Bargain? America and the Dream of Total Information Awareness."

11. Marx, "Technology and Social Control."

12. Lyon, *Surveillance after September 11.*

13. Wood, Konvitz, and Ball, "The Constant State of Emergency? Surveillance after 9/11."

14. Lyon, *Surveillance Society.*

15. Simmons, "Why 2007 Is Not Like 1984."

16. Goodin, "NSA Repeatedly Tries to Unpeel Tor Anonymity and Spy on Users, Memos Show."

17. Haggerty and Ericson, "New Politics of Surveillance and Visibility."

18. Deleuze, "Postscript on the Societies of Control."

19. Norris and Armstrong, *Maximum Surveillance Society.*

20. Ceyhan, "Technologization of Security"; Amoore and de Goede, "Governance, Risk and Dataveillance in the War on Terror."

21. Ball and Webster, "Intensification of Surveillance."

22. Bogard, "Welcome to the Society of Control."

23. Monahan, "Electronic Fortification in Phoenix Surveillance Technologies and Social Regulation in Residential Communities."

24. Polgreen, "With National Database, India Tries to Reach the Poor."

25. Leo and Richman, "Mandate the Electronic Recording of Police Interrogations."

26. Sweet and Cass, "How to Fight Crime in Real Time."

27. Bogard, "Welcome to the Society of Control."

28. Jang, Hoover, and Joo, "Evaluation of Compstat's Effect on Crime."

29. Markon and Nakashima, "NSA Director Says Surveillance Programs Thwarted 'Dozens' of Attacks."

30. Guzik, "Discrimination by Design."

31. Aas, Gundhus, and Lomell, *Technologies of InSecurity;* Manning, *Technology of Policing;* Willis, Mastrofski, and Weisburd, "Making Sense of COMPSTAT"; Silverman, "Compstat's Innovation."

32. Brodeaur and Leman-Langlois, "Surveillance Fiction or Higher Policing?"

33. Gates, "Identifying the 9/11 'Faces of Terror.'"

34. Lee and Schwartz, "Beyond the 'War' on Terrorism."

35. Brown and Duguid, *Social Life of Information.*

36. Bornstein, "Antiterrorist Policing in New York City after 9/11"; Lyon, *Surveillance after September 11.*

37. Norris, "Success of Failure."

38. Cratty, "FBI Uses Drones in U.S., Says Mueller."

39. Replogle, "Senate Immigration Bill Calls for a Drone-Patrolled Border."

40. Monahan, *Surveillance in the Time of Insecurity,* 4; Gates, *Our Biometric Future.*

41. See Monahan, *Surveillance in the Time of Insecurity.*

42. Simon, *Governing through Crime.*

43. See Braverman, "Governing with Clean Hands"; and Braverman, "Civilized Borders."

44. Haggerty and Ericson, "Surveillant Assemblage"; Haggerty, "From Risk to Precaution."

45. Norris and Armstrong, *Maximum Surveillance Society.*

46. Ibid.

47. Cole and Lobel, *Less Safe, Less Free.*

48. Amoore and de Goede, "Governance, Risk and Dataveillance in the War on Terror"; Lyon, *Surveillance after September 11;* Lyon, *Surveillance Society.*

49. Sparks, "Fast Capitalism/Slow Terror"; Amoore and de Goede, "Governance, Risk and Dataveillance in the War on Terror." See also Braverman, "Civilized Borders."

50. Smith, "Exploring Relations between Watchers and Watched in Control(led) Systems"; Monahan, "'War Rooms' of the Street."

51. Feeley and Simon, "New Penology."

52. Goold, "Technologies of Surveillance and the Erosion of Institutional Trust."

53. See Electronic Frontier Foundation, www.eff.org; and Electronic Privacy Information Center, www.epic.org.

54. Marx, "Surveillance and Society."

55. Gilliom, "Struggling with Surveillance."

56. Graham and Wood, "Digitizing Surveillance."

57. Smith, "Exploring Relations between Watchers and Watched in Control(led) Systems."

58. Chrisafis, "NSA Surveillance."

59. Fortin, "Nigerian Citizens Call for Transparency amid Suspicions of Shady Government Deal for Internet Surveillance."

60. Polgreen, "With National Database, India Tries to Reach the Poor."

61. Hookway, "Thailand's Dummy Cops Are Really Watching This Time."

62. Wilson, "Road Pricing."

63. Doyle, Lippert, and Lyon, *Eyes Everywhere.*

64. Murakami Wood, "Globalization and Surveillance."

65. Murakami Wood, "Cameras in Context."

66. Firmino, Bruno, and Botello, "Understanding the Sociotechnical Networks of Surveillance Practices in Latin America."

67. Davis, "Age of Insecurity"; Arteaga Botello, "Surveillance Studies."

68. Pereira and Davis, "New Patterns of Militarized Violence and Coercion in the Americas."

69. Arias and Goldstein, "Violent Pluralism"; Toro and Serrano, "From Drug Trafficking to Transnational Organized Crime in Latin America."

70. Arias and Goldstein, "Violent Pluralism."

71. Ibid.

72. O'Donnell, "Reflections on Contemporary South American Democracies."

73. Freeman and Sierra, "Mexico: The Militarization Trap."

74. De la Luz González and Mejía, "Cae 'La Barbie' cerca del DF."

75. Hickey, *Encyclopedia of Murder and Violent Crime.*

76. Benítez Manaut, "Containing Armed Groups, Drug Trafficking, and Organized Crime in Mexico."

77. García, "Alertan por niveles de impunidad en México."

78. Instituto Nacional de Estadística y Geografía, "Encuesta nacional de victimización y percepción sobre seguridad pública 2014."

79. Ibid.

80. Consulta Mitofsky, *México: Confianza en instituciones.*

81. Guzik, "Security a la Mexicana."

82. Michel and Gómez, "Avanza ley contra secuestro"; Gómez and Michel, "Comisiones del Senado avalan Ley de Seguridad Nacional."

83. Sabet, *Police Reform in Mexico.*

84. Cook, Rush, and Ribando Seelke, *Merida Initiative.*

85. *BBC Mundo,* "México: Entre la responsabilidad y la censura de los medios."

86. Ávila Pérez, "Margarita Zavala."

210 | Notes

87. Cook, Rush, and Ribando Seelke, *Merida Initiative*.

88. Thompson and Mazzetti, "U.S. Sends Drones to Fight Mexican Drug Trade."

89. Yin, *Case Study Research*.

90. See Arias and Goldstein, *Violent Democracies in Latin America*; and Guzik, "Security a la Mexicana."

91. Rubin and Babbie, *Essential Research Methods for Social Work*.

92. Scott, *Seeing Like a State*.

93. Aas, *Sentencing in the Age of Information*.

94. Gilliom and Monahan, *SuperVision*.

95. Lyon, *Identifying Citizens*.

96. Haggerty, "Visible War."

97. Barrett, "One Surveillance Camera for Every 11 People in Britain, Says CCTV Survey."

98. Mann, Nolan, and Wellman, "Sousveillance."

99. Baudrillard, *Simulacra and Simulation*, 111, 141.

100. Bogard, *Simulation of Surveillance*, 21.

101. Clarke, "Information Technology and Dataveillance."

102. Marres and Lezaun, "Materials and Devices of the Public"; Braun and Whatmore, *Political Matter*.

103. Foucault, *Discipline and Punish*, 10.

104. For research on surveillance and the body, see Littlefield, "Constructing the Organ of Deceit"; Epstein, "Embodying Risk"; Monahan and Wall, "Somatic Surveillance"; Bogard, "Welcome to the Society of Control"; Lyon, *Surveillance Society*; and Haggerty and Ericson, "Surveillant Assemblage."

105. Moore, *Criminal Artefacts Governing Drugs and Users*.

106. Amoore, *Politics of Possibility*.

107. Gates, "Biometrics and Post-9/11 Technostalgia."

108. For such other works, see Marx, *Windows into the Soul*; Manning, *Technology of Policing*; and Breckenridge, "Elusive Panopticon."

109. Marx, *Windows into the Soul*, chapter 14.

110. Carroll, *Science, Culture, and Modern State Formation*.

111. Jones, *Desert Kingdom How Oil and Water Forged Modern Saudi Arabia*.

112. Barry, *Material Politics*.

113. Medina, *Cybernetic Revolutionaries Technology and Politics in Allende's Chile*.

114. Tilly, *Coercion, Capital, and European States, AD 990–1992*.

115. Jasanoff, *States of Knowledge*. See also Braun and Whatmore, *Political Matter*; and Marres and Lezaun, "Materials and Devices of the Public."

116. DeLanda, *New Philosophy of Society*; Anderson et al., "On Assemblages and Geography."

117. See Latour, *Pandora's Hope*; and Pickering, *Mangle of Practice*.

118. See Deleuze and Guattari, *Thousand Plateaus*; DeLanda, *Philosophy and Simulation*.

119. Lévi-Strauss, *Savage Mind*.

120. Scott, *Seeing Like a State*.

121. Velasco, *Insurgency, Authoritarianism, and Drug Trafficking in Mexico's "Democratization"*; Moser, "Editor's Introduction: Urban Violence and Insecurity."

122. Arias and Goldstein, *Violent Democracies in Latin America;* Serrano and Toro, "Del narcotráfico al crimen transnacional organizado en América Latina."

123. Guzik, "Security a la Mexicana."

124. Arteaga Botello, "Video-vigilancia del espacio urbano; Vélez, "Insecure Identities."

125. Anderson et al., "On Assemblages and Geography."

CHAPTER 2. TAMING THE TIGER

1. Secretaría de Gobernación (SEGOB), "Inicio del registro para obtener la Cédula de Identidad para Menores."

2. Arteaga Botello and Fuentes Rionda, "Nueva lógica de la seguridad en México."

3. *Milenio,* "Inauguran centro de mando en Huixquilucan."

4. Arteaga Botello, "Video-vigilancia del espacio urbano."

5. Cook, Rush, and Ribando Seelke, *Merida Initiative.*

6. Ribando Seelke and Finklea, *U.S.-Mexican Security Cooperation.*

7. Thompson and Mazzetti, "U.S. Sends Drones to Fight Mexican Drug Trade."

8. Meserve and Ahlers, "Drone Crash in El Paso under Investigation."

9. Thompson and Mazzetti, "U.S. Sends Drones to Fight Mexican Drug Trade."

10. Jasanoff, *States of Knowledge.*

11. Bennett, *Vibrant Matter.*

12. *El Universal,* "Promulgan registro nacional de celulares."

13. Secretaría de Comunicaciones y Transportes, "Decreto por el que se reforman y adicionan diversas disposiciones de la Ley Federal de Telecomunicaciones"; Comisión Federal de Telecomunicaciones, "Reglas del registro nacional de usuarios de telefonía móvil."

14. *El Universal,* "Promulgan registro nacional de celulares."

15. Bolaños, "Hoy, el último adiós a Silvia Vargas Escalera."

16. Duarte, "Secuestro de Martí conmociona a México."

17. Martínez and Morales, "Exige Martí que gobiernos se coordinen contra secuestro."

18. Secretaría de Gobernación (SEGOB), "Preguntas frecuentes."

19. *El Universal,* "Promulgan registro nacional de celulares."

20. Registro Nacional de Población (RENAPO), "¿Qué es la Cédula de Identidad Personal (registro de menores de edad)?"

21. *El Universal,* "Entérate ¿Para qué servirá la Cédula de Identidad?"

22. Transparencia Mexicana, "Informe."

23. Mejía, "Empresa extranjera hará cédula."

24. *El siglo de Torreón,* "Arranca el registro público de vehículos"; Secretaría de Seguridad Pública, "Sobre REPUVE."

25. *El Siglo de Torreón,* "Arranca el registro público de vehículos."

26. De la Luz González, "Obligatorio, el chip vehicular, dice SSP."

27. *El Siglo de Torreón,* "Inicia Calderón el programa de identificación vehicular."

28. Ponce, "Inicia a nivel nacional el Registro Público Vehicular en Zacatecas."

29. Nunn, "Cities, Space, and the New World of Urban Law Enforcement Technologies."

30. Manning, *Technology of Policing.*

31. Marx, "Surveillance and Society."

32. Haggerty and Ericson, "New Politics of Surveillance and Visibility"; Amoore and de Goede, "Governance, Risk and Dataveillance in the War on Terror."

33. Sparks, "Fast Capitalism/Slow Terror"

34. Amoore and de Goede, "Governance, Risk and Dataveillance in the War on Terror."

35. Foucault, *Security, Territory, Population,* 352–55.

36. Lyon, *Surveillance after September 11.*

37. See Arteaga Botello, "Video-vigilancia del espacio urbano."

38. See Murakami Wood, "Globalization and Surveillance."

39. Castillo, "A la fosa común, 97% de los cuerpos no identificados en la guerra antinarco de Calderón."

40. Van Dijk, van Kesteren, and Smit, *Criminal Victimisation in International Perspective.*

41. Scott, *Seeing Like a State,* 1.

42. Ibid., 185.

43. Ibid., 72.

44. Ibid., 65.

45. Ling, *Mobile Connection;* Geser, "Is the Cell Phone Undermining the Social Order?"

46. Urry, "'System'of Automobility."

47. Lyon, *Identifying Citizens,* 611.

48. Callon, "Some Elements of a Sociology of Translation."

49. Pickering, *Mangle of Practice,* 170.

50. Díaz del Castillo, *Memoirs of the Conquistador Bernal Diaz del Castillo.*

51. Díaz del Castillo, *Historia verdadera de la conquista de la Nueva España,* 85.

52. Ibid., 220.

53. García Martinez, "Los años de la conquista," 183; MacLachlan, *Criminal Justice in Eighteenth Century Mexico,* 14–15.

54. García Martinez, "Los años de la conquista," 179.

55. Cope, *Limits of Racial Domination.*

56. García Martinez, "Los años de la conquista," 193–95; MacLachlan, *Criminal Justice in Eighteenth Century Mexico,* 27–28.

57. Cope, *Limits of Racial Domination.*

58. Carrera, "Locating Race in Late Colonial Mexico."

59. Cope, *Limits of Racial Domination,* 16.

60. Hausberger and Mazín, "Nueva España: Los años de autonomía," 292.

61. Ibid.

62. Cope, *Limits of Racial Domination.*

63. Ibid., 16.

64. Ibid., 18.

65. Ibid.

66. Carrera, *Imagining Identity in New Spain.*

67. Vanderwood, *Disorder and Progress: Bandits, Police, and Mexican Development,* 15.

68. García Martinez, "Los Años de Expansión."

69. Vanderwood, *Disorder and Progress.*

70. García Martinez, "Los años de expansión."

71. Vanderwood, *Disorder and Progress,* 18.

72. MacLachlan, *Criminal Justice in Eighteenth Century Mexico.*

73. Callon, "Some Elements of a Sociology of Translation."

74. Serrano Ortega and Zoraida Vázquez, "El nuevo orden"; Buffington, "Periodization and Its Discontents," 98.

75. Buffington, *Criminal and Citizen in Modern Mexico.*

76. Ibid.; Buffington, "Periodization and Its Discontents."

77. Warren, "Mass Mobilization versus Social Control," 44.

78. Ibid.

79. Jusidman, "El padrón electoral en el camino de la democracia en México."

80. Warren, "Mass Mobilization versus Social Control"; Buffington, *Criminal and Citizen in Modern Mexico.*

81. Jusidman, "El padrón electoral en el camino de la democracia en México."

82. Kuntz Ficker and Speckman Guerra, "El Porfiriato."

83. Van Hoy, *Social History of Mexico's Railroads.*

84. Van Hoy, "La Marcha Violenta?," 9.

85. Ibid.

86. See also Müller, *Public Security in the Negotiated State,* for other instances of the central role of local strongmen in the development of modern Mexico.

87. Van Hoy, "La Marcha Violenta?," 51.

88. Van Hoy, *Social History of Mexico's Railroads,* 209.

89. Ibid., n.p.

90. Aboites and Loyo, "La construcción del nuevo estado"; Gonzales, *Mexican Revolution, 1910–1940;* Hernández Chávez, "El estado nacionalista, su referente histórico"; Matute Aguirre, "La encrucijada de 1929."

91. See Joseph and Nugent, "Popular Culture and State Formation in Revolutionary Mexico," 7.

92. Ervin, "Statistics, Maps, and Legibility."

93. Ibid.

94. Ibid., 162.

95. Ervin, "1930 Agrarian Census in Mexico"; Ervin, "Statistics, Maps, and Legibility."

96. Hernández Serrano, *. . . y se formaron caminos*, 36.

97. García Martínez, *The Highway of Mexico (1891–1991)*, 30.

98. Ibid., 34.

99. Ibid.

100. Hernández Serrano, *. . . y se formaron caminos*, 140.

101. Ibid., 146.

102. Ibid.; García Martínez, *Highway of Mexico*, 42.

103. Ibid., 56.

104. Ibid., n.p.

105. Ibid.

106. Jusidman, "El padrón electoral en el camino de la democracia en México."

107. Ibid.

108. Campos and Penna, *Confianza en las instituciones*.

109. Carroll, *Science, Culture, and Modern State Formation*; Passoth and Rowland, "Actor-Network State."

110. Jasanoff, *States of Knowledge*, 14.

111. Hausberger and Mazín, "Nueva España."

112. Tanck de Estrada and Marichal, "¿Reino o Colonia?."

113. Bennett, *Vibrant Matter*.

114. Carroll, "Hillary Clinton."

115. Koonings, "New Violence, Insecurity, and the State."

116. Márquez and Meyer, "Del autoritarismo agotado a la democracia frágil."

117. CONEVAL (Consejo Nacional de Evaluación de la Política de Desarrollo Social), "Evolución de las dimensiones de la pobreza, 1990–2010."

118. Dresser, "Mexico: From PRI Predominance to Divided Democracy."

119. Serrano, "States of Violence."

120. Levy, Bruhn, and Zebadúa, *Mexico: The Struggle for Democratic Development*, 273–75; Hernández, *Narcoland*.

121. Rubenstein, "Mass Media and Popular Culture in the Postrevolutionary Era."

122. Shirk and Cázares, "Introduction: Reforming the Administration of Justice in Mexico," 19.

123. Nelson Reames, "Profile of Police Forces in Mexico."

124. Ervin, "1930 Agrarian Census in Mexico," 537.

125. Latour, "Where Are the Missing Masses?"

126. Ibid.

CHAPTER 3. PROHESION

1. Recillas Enecoiz, *"El automóvil gris."* Quotations from the film are all my translations.

2. Navitski, "Spectacles of Violence and Politics," 136.

3. *La Cultura Jurídica*, "El orden constitucional y la banda del *Automovil gris*."

4. Navitski, "Spectacles of Violence and Politics."

5. Román, "Robo de autos creció 86% en seis años."

6. *El Informador,* "Delincuentes bloquean vías carreteras a Miguel Alemán y a Reynosa."

7. De la Luz González, "Coche-bomba mata a 3 en Juárez."

8. Urry, "'System'of Automobility."

9. Asociación Mexicana de Distribuidores de Automotores, *Un siglo en movimiento.*

10. Fernández Christlieb, *Modernas ruedas de la destrucción,* 18.

11. Ibid., 33.

12. Fernández Christlieb, *Modernas ruedas de la destrucción.*

13. Asociación Mexicana de Distribuidores de Automotores, *Un siglo en movimiento,* 17–19.

14. Plana, *Industrias, siglos XVI al XX,* 108; Zegarra Ballón T, *Censo de automóviles en América Latina,* 61.

15. Zegarra Ballón T, *Censo de automóviles en América Latina,* 61.

16. Fernández Christlieb, *Modernas ruedas de la destrucción,* 34.

17. Zegarra Ballón T, *Censo de automóviles en América Latina,* 61–64.

18. Ibid., 59–61.

19. Urry, "'System'of Automobility," 25.

20. Asociación Mexicana de Distribuidores de Automotores, *Un siglo en movimiento.*

21. Ramírez de la O, "Impact of NAFTA on the Auto Industry in Mexico."

22. Zegarra Ballón T, *Censo de automóviles en América Latina,* 65.

23. Fernández Christlieb, *Modernas ruedas de la destrucción,* 35.

24. Ibid., 34.

25. Asociación Mexicana de Distribuidores de Automotores, *Un siglo en movimiento;* Roxborough, *Unions and Politics in Mexico.*

26. Zegarra Ballón T, *Censo de automóviles en América Latina,* 9; Womack, *Mexican Motor Industry,* 110–12.

27. Asociación Mexicana de Distribuidores de Automotores, *Un siglo en movimiento,* 34.

28. Ibid., 49–50.

29. Asociación Mexicana de Distribuidores de Automotores, *Un siglo en movimiento,* 66–68.

30. Womack, *Mexican Motor Industry,* 112.

31. Asociación Mexicana de Distribuidores de Automotores, *Un siglo en movimiento,* 73–75.

32. Ramírez de la O, "Impact of NAFTA on the Auto Industry in Mexico," 56.

33. Plana, *Industrias, siglos XVI al XX.*

34. Womack, *Mexican Motor Industry,* 107–9.

35. Asociación Mexicana de Distribuidores de Automotores, *Un siglo en movimiento,* 105.

36. *Economist,* "Steaming Hot."

37. Case, "Mexico Surpassing Japan as No. 2 Auto Exporter to U.S."

38. Fernández Christlieb, *Modernas ruedas de la destrucción,* 108–10.

39. Ibid., n.p.

40. Ibid., 56–8.

41. Ibid., 56.

42. Ibid., 111.

43. Ibid., n.p.

44. Ibid., 17.

45. Urry, "'System'of Automobility," 25.

46. Fernández Christlieb, *Modernas ruedas de la destrucción*, 37, 99.

47. Simon, "Driving Governmentality."

48. McShane, "Origins and Globalization of Traffic Control Signals."

49. Sanger, "Girls and the Getaway."

50. Simon, *Governing through Crime*.

51. Gusfield, *Culture of Public Problems*.

52. Secretaría de Comunicaciones y Obras Públicas (SCOP), *Seguridad!*.

53. World Health Organization, *Global Status Report on Road Safety*.

54. Zegarra Ballón T, *Censo de automóviles en América Latina*.

55. World Health Organization, *Global Status Report on Road Safety*.

56. Secretaría de Comunicaciones y Obras Públicas (SCOP), *Seguridad!*, 7.

57. Ibid., n.p.

58. Ibid., 29.

59. Ibid.

60. Ibid.

61. Sanchiz, *Manual del "chauffeur,"* 322.

62. Ayuntamiento de Ciudad Juárez, *Reglamento de vehículos para la municipalidad de C. Juárez*, 11.

63. Sanchiz, *Manual del "chauffeur,"* 322.

64. Ayuntamiento de Ciudad Juárez, *Reglamento de vehículos para la municipalidad de C. Juárez*, 10.

65. Secretaría de Comunicaciones y Obras Públicas (SCOP), *Seguridad!*, 25.

66. Sanchiz, *Manual del "chauffeur,"* 321.

67. Ibid., 322.

68. Secretaría de Comunicaciones y Obras Públicas (SCOP), *Seguridad!*, 28.

69. Sanchiz, *Manual del "chauffeur,"* 311.

70. Ayuntamiento de Ciudad Juárez, *Reglamento de vehículos para la municipalidad de C. Juárez*, 6.

71. Ibid., 7.

72. Ibid., 11.

73. Ibid., 10.

74. Ibid.

75. Ibid., 11, 22.

76. Ibid., 20–21.

77. Rose, O'Malley, and Valverde, "Governmentality."

78. Scott, *Seeing Like a State*.

79. Foucault, *Discipline and Punish*, 184.

80. Simon, "Driving Governmentality," 555–56.

81. Valverde, "Police Science, British Style."

82. Fernández Christlieb, *Modernas ruedas de la destrucción*, 45–63.

83. Beck, *Risk Society*, 21.

84. Centro de Estudios del Sector Privado para el Desarrollo Sustenable (CESPEDES), "Normatividad amiental y emisiones vehiculares en México," 33.

85. Asociación Mexicana de Distribuidores de Automotores, *Un siglo en movimiento*, 70.

86. Secretaría de Medio Ambiente, *Elementos para la propuesta de actualización del programa "Hoy No Circula."*

87. Fernández Christlieb, *Modernas ruedas de la destrucción.*

88. O'Connor, "Mexico City Drastically Reduced Air Pollutants since 1990s."

89. Beck, *Risk Society.*

90. Latour, "Give Me a Lab and I Will Raise the World."

91. Ibid.

92. Román, "Robo de autos creció 86% en seis años."

93. De la Luz González, "Coche-bomba mata a 3 en Juárez."

94. Davis, "Policing and Regime Transition:."

95. *El Espectador,* "Adiós al 'vocho', el taxi del pueblo."

96. Ley del Registro Público Vehicular, 1.

97. Ibid., 3.

98. El Secretariado Ejecutivo del Sistema Nacional de Seguridad Pública, *Libro Blanco*, 23.

99. Ibid., 26; Castro Medina, *Criminalística en la identificación de vehículos automotores,* 20; Asociación Mexicana de Distribuidores de Automotores, *Un siglo en movimiento,* 104.

100. El Secretariado Ejecutivo del Sistema Nacional de Seguridad Pública, *Libro Blanco*, 26.

101. Lutz and Roht-Arriaza, "Cavallo Case"; El Secretariado Ejecutivo del Sistema Nacional de Seguridad Pública, *Libro Blanco.*

102. Lutz and Roht-Arriaza, "Cavallo Case."

103. Castro, "Perfil de Ricardo Miguel Cavallo."

104. El Secretariado Ejecutivo del Sistema Nacional de Seguridad Pública, *Libro Blanco*, 27.

105. Ley del Registro Público Vehicular, 3.

106. El Secretariado Ejecutivo del Sistema Nacional de Seguridad Pública, *Libro Blanco*, 1.

107. Ibid., 20–21.

108. Ibid., 21.

109. Ibid.

110. Ibid., 2.

111. Ibid., 21.

112. Ibid., 28.

113. El Secretariado Ejecutivo del Sistema Nacional de Seguridad Pública, *Libro Blanco*, 55–56.

114. Ley del Registro Público Vehicular, 3.

115. Cámara de Diputados del H. Congreso de la Unión, "Reglamento de la Ley del Registro Público Vehicular," 2.

116. Ibid., 9.

117. Ley del Registro Público Vehicular, 6–7.

118. Cámara de Diputados del H. Congreso de la Unión, "Reglamento de la Ley del Registro Público Vehicular," 2.

119. El Secretariado Ejecutivo del Sistema Nacional de Seguridad Pública, *Libro Blanco*, 1.

120. Ley del Registro Público Vehicular, 3.

121. El Secretariado Ejecutivo del Sistema Nacional de Seguridad Pública, *Libro Blanco*, 59–60.

122. Cámara de Diputados del H. Congreso de la Unión, "Reglamento de la Ley del Registro Público Vehicular," 9.

123. El Secretariado Ejecutivo del Sistema Nacional de Seguridad Pública, *Libro Blanco*, 7.

124. Ibid., 6.

125. Ibid., 7–8, 66–67.

126. Ibid., 64.

127. Ibid., 65.

128. Ibid., 66–67.

129. Ibid., 66–68.

130. Prado, "Presume SSP su plataforma México."

131. Lyon, *Surveillance after September 11*, 8–10.

132. Jasanoff, *States of Knowledge*.

133. Virilio, *Information Bomb*.

134. Nellis, "Mobility, Locatability and the Satellite Tracking of Offenders," 107.

135. Ibid., n.p.

136. Latour, "Where Are the Missing Masses?"

137. Clarke, "Information Technology and Dataveillance," 498.

138. Foucault, *Discipline and Punish*.

139. Poster, *Mode of Information*, 93.

140. Marx, "Surveillance and Society," 2.

141. Haggerty and Ericson, "Surveillant Assemblage," 605.

142. Lyon, *Surveillance Society*, 3.

143. *Online Etymology Dictionary*, www.etymonline.com.

144. Hacking, *Taming of Chance*.

145. Foucault, *Discipline and Punish*.

146. Deleuze, "Postscript on the Societies of Control"; Rose, *Powers of Freedom*.

147. Secretaría de Comunicaciones y Transportes, "Decreto por el que se reforman y adicionan diversas disposiciones de la Ley Federal de Telecomunicaciones"; Comisión Federal de Telecomunicaciones, "Reglas del registro nacional de usuarios de telefonía móvil."

148. Secretaría de Gobernación, "Preguntas frecuentes."

149. Registro Nacional de Población (RENAPO), "¿Qué es la Cédula de Identidad Personal (registro de menores de edad)?"

CHAPTER 4. NI CON GOMA

1. *El Universal*, "Promulgan registro nacional de celulares."

2. Secretaría de Gobernación (SEGOB), "Preguntas frecuentes."

3. Vargas, "Privacy Rights under Mexican Law."

4. Notimex, "Titular de Cofetel defiende 'éxito' de Renaut."

5. Posada García, "Movistar perderá la concesión si incumple la ley, advierte Cofetel."

6. Soliloquio21, "Trucos para evadir RENAUT."

7. El Pop, "Las fallas del CURP y como registrar anonimamente tu celular en el RENAUT."

8. Monroy, "Extorsiones aumentan 40% por Renaut."

9. Olivares Alonso, "Concretar la Cédula de Identidad significa echar a la basura $40 mil millones, advierten."

10. Campos and Penna, *Confianza en las instituciones.*

11. *El Siglo de Torreón,* "Inicia Calderón el programa de identificación vehicular."

12. Scott, *Weapons of the Weak.*

13. Ewick and Silbey, *Common Place of Law.*

14. Foucault, *Discipline and Punish,* 95.

15. Scott, *Weapons of the Weak,* 29.

16. Olson, *Logic of Collective Action.*

17. Jenkins, "Resource Mobilization Theory and the Study of Social Movements."

18. McAdam, McCarthy, and Zald, *Comparative Perspectives on Social Movements.*

19. Touraine, "Introduction to the Study of Social Movements."

20. Melucci, *Nomads of the Present.*

21. Benford and Snow, "Framing Processes and Social Movements."

22. Goodwin, Jasper, and Polletta, *Passionate Politics.*

23. Polletta, "'It Was Like a Fever.'"

24. Barclay, Bernstein, and Marshall, *Queer Mobilizations.*

25. McCann, *Rights at Work.*

26. Scott, *Weapons of the Weak,* 290–96.

27. Ibid., 296.

28. Ibid., n.p.

29. Merry, "Resistance and the Cultural Power of Law."

30. Merry, "Resistance and the Cultural Power of Law," 15.

31. Zureik et al., *Surveillance, Privacy, and the Globalization of Personal Information.*

32. Morales Zebadúa, "Breve historia de pifias en la protección de los datos personales en México."

33. Olvera, "Población desconfía de los registros."

34. *El Diario de Yucatán,* "La Cédula de Identidad, a debate."

35. Villanueva, "Mi columna semanal."

36. Scott, *Seeing Like a State,* 299–300.

37. Guadarrama, "El Renaut dio como resultado un fracaso millonario."

38. Organisation for Economic Co-operation and Development, *Collusion and Corruption in Public Procurement.*

39. Chacón, "Advierten opacidad en Repuve."

40. Ibid.

41. Raphael, "El chip de Campa."

42. De la Luz González, "Federación y los estados compartirán gastos."

43. Ewick and Silbey, *Common Place of Law.*

44. Ibid.

45. Ramírez, "Inoperate el registro vehicular."

46. Mendoza, "Comparecen los titulares de SCT, Cofetel, y Cofeco ante Comisión de Comunicación."

47. Müller, *Public Security in the Negotiated State;* Pansters, "Zones of State-Making"; Van Hoy, *Social History of Mexico's Railroads.*

48. Funetes, "Buscan poseer la marca Repuve."

49. García Martinez, "Los años de expansión."

50. Arias, *Drugs and Democracy in Rio de Janeiro.*

51. Powell, "Political Violence, Everyday Political Violence, and Electoral Processes during the Neoliberal Period in Mexico."

52. Morris, *Political Corruption in Mexico.*

53. Müller, *Public Security in the Negotiated State;* Pansters, "Zones of State-Making; Van Hoy, *Social History of Mexico's Railroads.*

54. Quiroz and Zepeda, "Gobernación para expedir la CIC obligó al gobierno federal a diferir el plan."

55. Márquez and Meyer, "Del autoritarismo agotado a la democracia frágil"; Morris, *Political Corruption in Mexico.*

56. Sánchez Ley, "30 millones con datos falsos en el Renaut."

57. *El Universal,* "Registro de celulares, sin mucho éxito: Cofetel."

58. *El Informador,* "Sin avance, comprobación de identidad de usuarios inscritos en Renaut."

59. Santillán, "20 mil vehículos dados de alta en el Repuve Tlaxcala."

60. Washington State Department of Transportation, "List of Vehicles with Metallized Windshields."

61. Ervin, "1930 Agrarian Census in Mexico."

62. Müller, *Public Security in the Negotiated State.*

63. Lin, "Silenced Technology"; Verrips and Meyer, "Kwaku's Car."

64. Callon, "Some Elements of a Sociology of Translation."

65. Latour, "Give Me a Lab and I Will Raise the World."

66. Pickering, *Mangle of Practice.*

67. Ibid.

68. Jasanoff, *States of Knowledge.*

69. Anderson et al., "On Assemblages and Geography."

70. Scott, *Weapons of the Weak,* 296.

71. Merry, "Resistance and the Cultural Power of Law," 25.

72. Molyneux, "Mobilization without Emancipation?"; Alvarez, *Engendering Democracy in Brazil.*

CHAPTER 5. STATECRAFT

1. *El Diario,* "Esperan hasta 30 horas por Repuve."

2. Herrera, "Se quedan a dormir para poder sacar engomado de REPUVE."

3. *El Diario,* " . . . Y acampan por Repuve."

4. *El Heraldo de Chihuahua,* "Reinicia actividad el 6 de enero Registro Público Vehicular."

5. Rodríguez, "Pernoctan en el parque por el Repuve."

6. *Sipse.com,* "Registro vehícular abre sus puertas los sábados."

7. Holguín, "Piden condonar los adeudos en 'plaqueo.'"

8. *Pulso,* "Policías ayudarán a compradores a verificar status de autos."

9. *ADN Sureste,* "SSPO equipa con tecnología de punta patrullas de Policía Vial."

10. Aguirre and Zepeda, "La Cédula de Identidad va, pero no por ahora."

11. *Milenio,* "Entrega Calderón primeras Cédulas de Identidad Personal para Menores."

12. Presidencia de la República, "Diversas intervenciones en el inicio del registro para obtener la Cédula de Identidad Personal."

13. Becerril and Ballinas, "El Senado pone fin al fallido registro de usuarios de celular."

14. Reyes, "InfoDF pide a la Segob eliminar datos del Renaut"; Medieta, "Exige el IFAI destruir los datos reunidos con el derogado Renaut."

15. *Informador,* "La Segob borra la base de datos del Renaut."

16. Aviña, "Piden cambios al Registro Público Vehicular."

17. Notimex, "Se alista el sector automotriz para negociar con nueva legislatura."

18. Herrera Madrano, "Circulan 10 mil autos robados en Tamaulipas."

19. Sánchez, "Arranca Registro Público Vehicular en Michoacán."

20. Notimex, "Aprueba Congreso de SLP que REPUVE sea obligatorio hasta 2014."

21. Martínez, "Instaló finanzas 8 mil chips del Repuve."

22. *Mundodehoy.com,* "Impuesto a coches importados."

23. López Galicia, "Imponen otro cobro a importadores de autos."

24. *Mundodehoy.com,* "Impuesto a coches importados."

25. López Galicia, "Imponen otro cobro a importadores de autos."

26. Juárez Alfaro, "Rechazan el Repuve."

27. Castro, "Denuncia la UIVAC cobro ilegal que hace banjército."

28. Quiroz and Zepeda, "Gobernación para expedir la CIC obligó al gobierno federal a diferir el plan."

29. Ibid.

30. *El Universal,* "Respalda Unicef proyecto de Cédula de Identidad Personal para Menores."

31. Campos and Penna, *Confianza en las instituciones.*

32. Notimex, "Detallan beneficios de la Cédula de Identidad personal."

33. Radial Sur, "'Libre tránsito' para gente del sur."

34. Becerril and Ballinas, "El Senado pone fin al fallido registro de usuarios de celular."

35. *El Informador,* "La Segob borra la base de datos del Renaut."

36. *Milenio,* "Telcel y Movistar desactivarán teléfonos celulares robados."

37. Hernández, "Facilitarán el bloqueo de celulares robados para frenar extorsión."

38. Guadarrama, "Renaut alterno ha dado de baja 12 mil 200 celulares."

39. Sánchez Ley, "Inútil, registro de celulares."

40. *Excelsior,* "Segob emite decreto de Ley de Geolocalización o monitoreo celular."

41. Juárez, "EFF considera 'alarmante' la nueva Ley de Geolocalización mexicana."

42. Ervin, "1930 Agrarian Census in Mexico."

43. Tilly, *Coercion, Capital, and European States, AD 990–1992;* Scott, *Seeing Like a State.*

44. Corrigan and Sayer, *Great Arch.*

45. Knight, "Modern Mexican State."

46. Jasanoff, *States of Knowledge;* Carroll, *Science, Culture, and Modern State Formation;* Passoth and Rowland, "Actor-Network State."

47. Anderson et al., "On Assemblages and Geography."

48. Tilly, *Coercion, Capital, and European States, AD 990–1992,* 25–26.

49. Knight, "Modern Mexican State," 192.

50. Müller, *Public Security in the Negotiated State.*

51. Flores, "La Ley de Geolocalización entra en vigor en México."

52. Rodríguez Manzo, "Geolocalización en el país de Las Maravillas."

53. *Milenio,* "Equivocada, iniciativa de Peña sobre telecom."

54. Zárate, "Desconocen beneficios de Cédula de Identidad."

55. Mejía, "La corte analiza queja por cédula."

56. Badillo, "La moral de los magnates Gates Y Slim."

57. Quiroz, "Cédula de Identidad sigue en pie."

58. Monroy, "Nuevo diseño de la credencial del IFE encriptará datos"; Nacif Hernández, "El INE frente al robo de identidad."

59. Fierro, "Llevará angulo fraude fiscal del REPUVE a la Cámara de Diputados."

60. Rebolledo, "Ilegal, cobro de Repuve por Banjército, acusan."

61. Veracruzanos.info, "Veracruz, primer lugar en registro vehicular."

CHAPTER 6. GRASPING SURVEILLANCE

1. Pacto por México, http://pactopormexico.org.

2. Buscaglia, "Mexico's Deadly Power Vacuum."

3. *El País,* "México reformista."

4. *Semanario ZETA,* "Los muertos de EPN."

5. Ortega, "Gendarmería Nacional será una policía cercana a la gente."

6. Valenzuela Zuñiga et al., *Technological Innovation for Security in Latin America.*

7. Pickler, "Experts Say Drone Strikes Appear in Bounds of US Law."

8. Steinhauer and Weisman, "U.S. Surveillance in Place since 9/11 Is Sharply Limited."

9. *USA Today,* "Obama Bans Some Military Equipment Sales to Police."

10. Electronic Frontier Foundation, "Mass Surveillance Technologies."

11. *Online Etymology Dictionary,* www.etymonline.com.

12. Clarke, "Information Technology and Dataveillance."

13. Foucault, *Discipline and Punish.*

14. Deleuze, "Postscript on the Societies of Control," 5.

15. Amoore and de Goede, "Governance, Risk and Dataveillance in the War on Terror"; Lyon, *Surveillance after September 11;* Lyon, *Surveillance Society.*

16. Sparks, "Fast Capitalism/Slow Terror"; Amoore and de Goede, "Governance, Risk and Dataveillance in the War on Terror"; Braverman, "Civilized Borders."

17. Norris and Armstrong, *Maximum Surveillance Society.*

18. Smith, "Exploring Relations between Watchers and Watched in Control(led) Systems"; Monahan, "'War Rooms' of the Street."

19. Feeley and Simon, "New Penology."

20. Bauman and Lyon, *Liquid Surveillance: A Conversation.*

21. Monahan, *Surveillance in the Time of Insecurity,* 23. On the culture of insecurity, see also Molotch, *Against Security.*

22. Valverde, "Targeted Governance and the Problem of Desire."

23. Goold, "Technologies of Surveillance and the Erosion of Institutional Trust," 211.

24. Lyon, "National IDs in a Global World: Surveillance, Security, and Citizenship"; Lyon, *Surveillance after September 11.*

25. Aas, *Sentencing in the Age of Information.*

26. Poiré Romero, "It Doesn't Come Easy—Mexico's Fight for Security."

27. Gilliom, *Overseers of the Poor;* Norris and Armstrong, *Maximum Surveillance Society.*

28. Foucault, *Discipline and Punish,* 141.

29. Poster, *Mode of Information.*

30. Ibid.; Norris and Armstrong, *Maximum Surveillance Society;* Haggerty and Ericson, "Surveillant Assemblage."

31. Epstein, "Embodying Risk," 185.

32. Amoore and de Goede, "Governance, Risk and Dataveillance in the War on Terror," 16.

33. Brodeaur and Leman-Langlois, "Surveillance Fiction or Higher Policing?," 196.

34. Haggerty and Ericson, "New Politics of Surveillance and Visibility," 17.

35. Gates, "Identifying the 9/11 'Faces of Terror,'" 431.

36. Breckenridge, "Elusive Panopticon," 47.

37. Bender and Bierman, "Russia Contacted FBI 'Multiple' Times on Concerns about Alleged Boston Marathon Bomber."

38. Keating, "No One in Europe Is Tougher on Terror Than France."

39. Glanz, Rotella, and Sanger, "In 2008 Mumbai Attacks, Piles of Spy Data, but an Uncompleted Puzzle."

40. Barker and Baker, "New York Officers' Killer, Adrift and Ill, Had a Plan."

41. Dresser, "Mexico: From PRI Predominance to Divided Democracy."

42. Benítez Manaut, "Containing Armed Groups, Drug Trafficking, and Organized Crime in Mexico."

43. Márquez and Meyer, "Del autoritarismo agotado a la democracia frágil."

44. CONEVAL (Consejo Nacional de Evaluación de la Política de Desarrollo Social), "Evolución de las dimensiones de la pobreza, 1990–2010."

45. Gorlier and Guzik, *La política de género en América Latina: Debates, teorías, metodologías y estudios de caso;* Guzik and Gorlier, "History in the Making."

46. Gilliom, "Struggling with Surveillance"; Gilliom, *Overseers of the Poor.*

47. Marx, "Seeing Hazily (But Not Darkly) through the Lens," 379. See also Marx, *Windows into the Soul.*

48. Pickering, *Mangle of Practice.*

49. Anderson et al., "On Assemblages and Geography."

50. Bennett, *Vibrant Matter.*

51. Neyland, "Mundane Terror and the Threat of Everyday Objects," 38. For things as the subject of surveillance, see also Murakami Wood, "What Is Global Surveillance?"

52. Müller, *Public Security in the Negotiated State.*

53. Jasanoff, *States of Knowledge.*

54. Lévi-Strauss, *Savage Mind.*

55. Pickering, *Mangle of Practice,* 22–23.

56. Scott, *Seeing like a State,* 6.

57. Sewell, "Theory of Structure."

58. Jasanoff, *States of Knowledge;* Carroll, *Science, Culture, and Modern State Formation.*

59. Marx, *Windows into the Soul.*

60. Marres and Lezaun, "Materials and Devices of the Public," 496.

61. Toobin, "The Solace of Oblivion."

62. Star-Ledger Editorial Board, "Track Domestic Abusers with GPS Devices."

63. Incháustegui Romer et al., "Violencia feminicida en México"; Cladem, "8 de marzo"; *El Universal,* "Investigan con tecnología de punto feminicidios en Juárez."

64. Lyon, *Surveillance Society,* 153.

65. Monahan, "Surveillance as Governance," 95.

66. Shallwani, "NYPD Unveil Two Cameras for Officers."

67. Lopez, "Why Police Should Wear Body Cameras—and Why They Shouldn't."

68. Chouza, "Organizaciones de periodistas crean un mapa de agresiones en México."

69. Villamil, "Atenco, Ibero y a primavera mexicana en 2012."

70. Culp-Ressler, "GOP Lawmaker Refuses to Support Abortion Bill."

71. *BBC News,* "Germany and Brazil in UN Spy Draft."

Bibliography

Aas, Katja Franko. *Sentencing in the Age of Information: From Faust to Macintosh*. London: Glasshouse Press, 2005.

Aas, Katja Franko, Helene Oppen Gundhus, and Heidi Mork Lomell, eds. *Technologies of InSecurity: The Surveillance of Everyday Life*. New York: Routledge-Cavendish, 2008.

Aboites, Luis, and Engracia Loyo. "La construcción del nuevo estado, 1920–1945." In *Nueva historia general de México,* edited by Erik Velásquez García et al., 595–652. Mexico City: Colegio de México, 2010.

ADN Sureste. "SSPO equipa con tecnología de punta patrullas de Policía Vial (13:16 H)." June 28, 2015.

Aguirre, Carlos, and Aurora Zepeda. "La Cédula de Identidad va, pero no por ahora." *Excelsior,* January 21, 2011.

Alvarez, Sonia E. *Engendering Democracy in Brazil*. Princeton, NJ: Princeton University Press, 1990.

Amoore, Louise. *The Politics of Possibility: Risk and Security beyond Probability*. Durham, NC: Duke University Press, 2013.

Amoore, Louise, and Marieke de Goede. "Governance, Risk and Dataveillance in the War on Terror." *Crime, Law and Social Change* 43, nos. 2–3 (April 2005): 149–73.

Anderson, Ben, Matthew Kearnes, Colin McFarlane, and Dan Swanton. "On Assemblages and Geography." *Dialogues in Human Geography* 2, no. 2 (2012): 171–89.

Arias, Enrique Desmond. *Drugs and Democracy in Rio de Janeiro: Trafficking, Social Networks, and Public Security*. Chapel Hill: University of North Carolina Press, 2006.

Arias, Enrique Desmond, and Daniel M. Goldstein, eds. *Violent Democracies in Latin America*. Durham, NC: Duke University Press, 2010.

———. "Violent Pluralism: Understanding the New Democracies of Latin America." In *Violent Democracies in Latin America,* edited by Enrique Desmond Arias and Daniel M. Goldstein, 1–34. Durham, NC: Duke University Press, 2010.

Arteaga Botello, Nelson. "Surveillance Studies: An Agenda for Latin America." *Surveillance and Society* 10, no. 1 (July 18, 2012): 5–17.

———. "Video-vigilancia del espacio urbano: Tránsito, seguridad y control social." *Andamios: Revista de investigación social* 7, no. 14 (2010).

Arteaga Botello, Nelson, and Roberto Fuentes Rionda. "Nueva lógica de la seguridad en México: Vigilancia y control de lo público y lo privado." *Revista Argentina de sociología* 7, nos. 12–13 (2009).

Asociación Mexicana de Distribuidores de Automotores. *Un siglo en movimiento: Crónica de la distribución de automotores en México.* Mexico City: Asociación Mexicana de Distribuidores de Automotores, 2000.

Ávila Pérez, Édgar. "Margarita Zavala: Miedo alimenta la impunidad." *El Universal,* November 9, 2009.

Aviña, Carlos. "Piden cambios al Registro Público Vehicular." *El Sol de México,* October 6, 2009.

Ayuntamiento de Ciudad Juárez. *Reglamento de vehículos para la municipalidad de C. Juárez, aprobado por el 6 de abril de 1918.* Ciudad Juárez: Talleres Tipográficos de "El Orden," 1922.

Badillo, Miguel. "La moral de los magnates Gates y Slim—abren expediente a Cédula de Identidad—investigan el gasto por 3 mil millones." *El Diario de Coahuila,* February 18, 2013.

Ball, Kirstie, and Frank Webster. "The Intensification of Surveillance." In *The Intensification of Surveillance: Crime, Terrorism and Warfare in the Information Age,* edited by Kirstie Ball and Frank Webster, 1–15. London: Pluto Press, 2003.

Barclay, Scott, Mary Bernstein, and Anna-Maria Marshall, eds. *Queer Mobilizations: LGBT Activists Confront the Law.* New York: NYU Press, 2009.

Barker, Kim, and Al Baker. "New York Officers' Killer, Adrift and Ill, Had a Plan." *New York Times,* December 21, 2014.

Barrett, David. "One Surveillance Camera for Every 11 People in Britain, Says CCTV Survey." *Telegraph,* July 10, 2013.

Barry, Andrew. *Material Politics: Disputes along the Pipeline.* RGS-IBG Book Series. Chichester, UK: Wiley Blackwell, 2013.

Baudrillard, Jean. *Simulacra and Simulation.* Ann Arbor: University of Michigan Press, 1994.

Bauman, Zygmunt, and David Lyon. *Liquid Surveillance: A Conversation.* Malden, MA: Polity, 2013.

BBC Mundo. "México: Entre la responsabilidad y la censura de los medios." March 25, 2011.

BBC News. "Germany and Brazil in UN Spy Draft." November 2, 2013.

Becerril, Andrea, and Victor Ballinas. "El Senado pone fin al fallido registro de usuarios de celular." *La Jornada,* April 30, 2011.

Beck, Ulrich. *Risk Society: Towards a New Modernity.* Thousand Oaks, CA: Sage, 1992.

Bender, Bryan, and Noah Bierman. "Russia Contacted FBI 'Multiple' Times on Concerns about Alleged Boston Marathon Bomber." *Boston Globe,* April 23, 2013.

Benford, Robert D., and David A. Snow. "Framing Processes and Social Movements: An Overview and Assessment." *Annual Review of Sociology* 26 (2000): 611–39.

Benítez Manaut, Raúl. "Containing Armed Groups, Drug Trafficking, and Organized Crime in Mexico." In *Organized Crime and Democratic Governability: Mexico and the U.S.-Mexican Borderlands,* edited by John Bailey and Roy Goodson, 126–58. Pittsburgh: University of Pittsburgh Press, 2001.

Bennett, Jane. *Vibrant Matter: A Political Ecology of Things.* Durham, NC: Duke University Press, 2010.

Bogard, William. *The Simulation of Surveillance: Hypercontrol in Telematic Societies.* Cambridge: Cambridge University Press, 1996.

———. "Welcome to the Society of Control: The Simulation of Surveillance Revisited." In *The New Politics of Surveillance and Visibility,* edited by Kevin D. Haggerty and Richard V. Ericson. Toronto: University of Toronto Press, 2006.

Bolaños, Claudia. "Hoy, el último adiós a Silvia Vargas Escalera." *El Universal,* December 13, 2008.

Bornstein, Avram. "Antiterrorist Policing in New York City after 9/11: Comparing Perspectives on a Complex Process." *Human Organization* 64, no. 1 (March 1, 2005): 52–61.

Braun, Bruce, and Sarah Whatmore, eds. *Political Matter: Technoscience, Democracy, and Public Life.* Minneapolis: University of Minnesota Press, 2010.

Braverman, Irus. "Civilized Borders: A Study of Israel's New Crossing Administration." *Antipode* 43, no. 2 (March 1, 2011): 264–95.

———. "Governing with Clean Hands: Automated Public Toilets and Sanitary Surveillance." *Surveillance and Society* 8, no. 1 (September 2010): 1–27.

Breckenridge, Keith. "The Elusive Panopticon: The HANIS Project and the Politics of Standards in South Africa." In *Playing the ID Card,* edited by Colin Bennette and David Lyon, 39–56. London: Routledge, 2008.

Brodeaur, Jean-Paul, and Stephane Leman-Langlois. "Surveillance Fiction or Higher Policing?" In *The New Politics of Surveillance and Visibility,* edited by Richard V. Ericson and Kevin D. Haggerty, 171–98. Toronto: University of Toronto Press, 2006.

Brown, John Seely, and Paul Duguid. *The Social Life of Information.* Boston: Harvard Business Review Press, 2000.

Buffington, Robert. *Criminal and Citizen in Modern Mexico.* Lincoln: University of Nebraska Press, 2000.

———. "Periodization and Its Discontents: The Social Construction of Crime and Criminality in Modern Mexico." USMEX 2003–04 Working Paper Series, Center for US-Mexican Studies, La Jolla, CA, 2003.

Buscaglia, Edgardo. "Mexico's Deadly Power Vacuum." *New York Times,* May 30, 2013.

Callon, Michel. "Some Elements of a Sociology of Translation: Domestication of the Scallops and the Fishermen of St. Brieuc Bay." *Power, Action, and Belief* 32 (1986): 196–223.

Campos, Roy, and Carlos Penna. *Confianza en las instituciones.* Consulta Mitofsky, 2010. www.opinamexico.org/opinion/ConfianzaInstituciones.pdf.

Carrera, Magali M. *Imagining Identity in New Spain: Race, Lineage, and the Colonial Body in Portraiture and Casta Paintings.* Austin: University of Texas Press, 2003.

———. "Locating Race in Late Colonial Mexico." *Art Journal* 57, no. 3 (1998): 36–45.

Carroll, Patrick. *Science, Culture, and Modern State Formation.* Berkeley: University of California Press, 2006.

Carroll, Rory. "Hillary Clinton: Mexican Drugs War Is Colombia-Style Insurgency." *Guardian,* September 9, 2010.

Case, Brendan. "Mexico Surpassing Japan as No. 2 Auto Exporter to U.S." *Bloomberg,* January 31, 2014.

Castillo, Gustavo. "A la fosa común, 97% de los cuerpos no identificados en la guerra antinarco de Calderón." *La Jornada,* January 2, 2013.

Castro, Aída. "Perfil de Ricardo Miguel Cavallo." *El Universal,* February 29, 2008.

Castro, Salvador. "Denuncia la UIVAC cobro ilegal que hace banjército." *NorteDigital,* May 8, 2013.

Castro Medina, Ana Luisa. *La criminalística en la identificación de vehículos automotores.* Mexico City: Porrúa, 2000.

Centro de Estudios del Sector Privado para el Desarrollo Sustentable (CESPEDES). *Normatividad amiental y emisiones vehiculares en México.* Mexico City: CESPEDES, 1998.

Ceyhan, Ayse. "Technologization of Security: Management of Uncertainty and Risk in the Age of Biometrics." *Surveillance and Society* 5, no. 2 (2008).

Chacón, Lilia. "Advierten opacidad en Repuve." *Reforma,* April 22, 2008.

Chouza, Paula. "Organizaciones de periodistas crean un mapa de agresiones en México." *El País,* January 24, 2015.

Chrisafis, Angelique. "NSA Surveillance: French Human Rights Groups Seek Judicial Investigation." *Guardian,* July 11, 2013.

Christie Digital. *Federal Police Intelligence Center of the Public Security Secretariat.* Accessed August 16, 2015. www.christiedigital.com/casestudies/2010/federal-police-intelligence-center-mexico-city-video-wall-solution-customer-story.pdf.

Ciborra, Claudio. *The Labyrinths of Information: Challenging the Wisdom of Systems.* Oxford: Oxford University Press, 2002.

Cladem. "8 de marzo: Mensajes de Bachelet y Ban Ki Moon." Accessed January 25, 2015. www.cladem.org/campanas/educacion-no-sexista/prensa/69-ens-otros-medios/404-8-de-marzo-mensajes-de-bachelet-y-ban-ki-moon.

Clarke, Roger. "Information Technology and Dataveillance." *Communications of the ACM* 31, no. 5 (May 1988): 498–512.

Cole, David, and Jules Lobel. *Less Safe, Less Free: Why America Is Losing the War on Terror.* New York: New Press, 2007.

Comisión Federal de Telecomunicaciones. "Reglas del registro nacional de usuarios de telefonía móvil." May 15, 2009. www.renaut.gob.mx/work/models/RENAUT/Resource/328/1/images/Reglas_del_RENAUT.pdf.

CONEVAL (Consejo Nacional de Evaluación de la Política de Desarrollo Social). "Evolución de las dimensiones de la pobreza, 1990–2010." July 29, 2011. www.coneval.gob.mx.

Consulta Mitofsky. *México: Confianza en instituciones.* 2014. http://consulta.mx/index.php/estudios-e-investigaciones/mexico-opina.

Cook, Colleen, Rebecca Rush, and Clare Ribando Seelke. *Merida Initiative: Proposed U.S. Anticrime and Counterdrug Assistance for Mexico and Central America.* CRS Report RS22837. Washington, DC: Congressional Research Service, March 18, 2008.

Cope, R. Douglas. *The Limits of Racial Domination: Plebeian Society in Colonial Mexico City, 1660–1720.* Madison: University of Wisconsin Press, 1994.

Corrigan, Philip, and Derek Sayer. *The Great Arch: English State Formation As Cultural Revolution.* Oxford: Blackwell, 1985.

Cratty, Carol. "FBI Uses Drones in U.S., Says Mueller." *CNN.com*, June 19, 2003.

Culp-Ressler, Tara. "GOP Lawmaker Refuses to Support Abortion Bill: 'It Does Nothing to Improve Women's Health.'" *ThinkProgress*, February 5, 2014.

Davis, Diane E. "The Age of Insecurity: Violence and Social Disorder in the New Latin America." *Latin American Research Review* 41, no. 1 (2006): 178–97.

———. "Policing and Regime Transition: From Postauthoritarianism to Populism to Neoliberalism." In *Violence, Coercion, and State-Making in Twentieth-Century Mexico: The Other Half of the Centaur,* edited by Wil Pansters, 68–90. Stanford, CA: Stanford University Press, 2012.

de la Luz González, María. "'Cerebro' tecnológico enfrentará criminales." *El Universal,* November 25, 2009.

———. "Coche-bomba mata a 3 en Juárez." *El Universal,* July 16, 2010.

———. "Eligen chip inútil contra robacoches." *El Universal,* March 29, 2008.

———. "Federación y los estados compartirán gastos: Campa." *El Universal,* March 29, 2008.

———. "Obligatorio, el chip vehicular, dice SSP." *El Universal,* May 25, 2009.

———. "Roberto Campa renuncia al SNSP." *El Universal,* September 2, 2008.

de la Luz González, María, and Gerardo Mejía. "Cae 'La Barbie' cerca del DF." *El Universal,* August 31, 2010.

DeLanda, Manuel. *A New Philosophy of Society: Assemblage Theory and Social Complexity.* Annotated ed. London: Bloomsbury Academic, 2006.

———. *Philosophy and Simulation: The Emergence of Synthetic Reason.* London: Continuum, 2011.

Deleuze, Gilles. "Postscript on the Societies of Control." *October* 59 (1992): 3–7.

Deleuze, Gilles, and Félix Guattari. *A Thousand Plateaus: Capitalism and Schizophrenia*. Minneapolis: University of Minnesota Press, 1987.

Díaz del Castillo, Bernal. *Historia verdadera de la conquista de la Nueva España*. Fernández Editores, S.A., 1632. www.antorcha.net/biblioteca_virtual/historia/bernal/indice.html.

———. *The Memoirs of the Conquistador Bernal Diaz del Castillo, Vol. 1 (of 2), Written by Himself Containing a True and Full Account of the Discovery and Conquest of Mexico and New Spain*. Translated by John Ingram Lockhart. 1844. www.gutenberg.org/ebooks/32474.

Doyle, Aaron, Randy Lippert, and David Lyon. *Eyes Everywhere: The Global Growth of Camera Surveillance*. New York: Routledge, 2011.

Dresser, Denise. "Mexico: From PRI Predominance to Divided Democracy." In *Constructing Democratic Governance in Latin America*, 2nd ed., edited by Jorge Dominguez and Michael Shifter, 321–50 Baltimore: Johns Hopkins University Press, 2003.

Duarte, Enrique. "Secuestro de Martí conmociona a México." *CNNExpansión. com*, August 4, 2008.

Economist. "Mexico's Car Industry: Steaming Hot." November 15, 2013.

El Diario. "Esperan hasta 30 horas por Repuve; temen sanciones." December 24, 2012.

———. ". . . Y acampan por Repuve." December 16, 2013.

El Diario de Yucatán. "La Cédula de Identidad, a debate." January 23, 2011.

Electronic Frontier Foundation. Accessed February 1, 2015. www.eff.org.

———. "Mass Surveillance Technologies." Accessed August 10, 2015. www.eff.org/issues/mass-surveillance-technologies.

Electronic Privacy Information Center. Accessed February 1, 2015. www.epic.org.

El Espectador. "Adiós al 'Vocho,' el taxi del pueblo." May 20, 2012.

El Heraldo de Chihuahua. "Reinicia actividad el 6 de enero Registro Público Vehicular." January 1, 2014.

El Informador. "Delincuentes bloquean vías carreteras a Miguel Alemán y a Reynosa." March 18, 2010.

———. "La Segob borra la base de datos del Renaut." May 13, 2013.

———. "Sin avance, comprobación de identidad de usuarios inscritos en Renaut." November 4, 2010.

Ellul, Jacques. "The Technological Order." *Technology and Culture* 3, no. 4 (1962): 394–421.

El País. "México reformista." September 13, 2013.

El Pop. "Las fallas del CURP y como registrar anonimamente tu celular en el RENAUT." *La Cofradía Digital*, March 25, 2010.

El Secretariado Ejecutivo del Sistema Nacional de Seguridad Pública. *Libro Blanco*. Mexico City: El Secretariado Ejecutivo del Sistema Nacional de Seguridad Pública, 2012. http://segob.gob.mx/work/models/SEGOB/Resource/1325/1/images/Registro_Publico_Vehicular.pdf.

El Siglo de Torreón. "Arranca el registro público de vehículos." March 4, 2008.

———. "Inicia Calderón el programa de identificación vehicular." June 23, 2009.

El Universal. "Entérate ¿Para qué servirá la Cédula de Identidad?" July 28, 2009.

———. "Investigan con tecnología de punto feminicidios en Juárez." June 9, 2003.

———. "Promulgan registro nacional de celulares." February 10, 2009.

———. "Registro de celulares, sin mucho éxito: Cofetel." January 17, 2011.

———. "Respalda Unicef proyecto de Cédula de Identidad Personal para Menores." January 13, 2011.

Epstein, Charlotte. "Embodying Risk: Using Biometrics to Protect the Borders." In *Risk and the War on Terror,* edited by Louise Amoore and Marieke de Goede, 178–93. London: Routledge, 2008.

Ervin, Michael A. "The 1930 Agrarian Census in Mexico: Agronomists, Middle Politics, and the Negotiation of Data Collection." *Hispanic American Historical Review* 87, no. 3 (2007): 537–70.

———. "Statistics, Maps, and Legibility: Negotiating Nationalism in Post-Revolutionary Mexico." *Americas* 66, no. 2 (2009): 155–79.

Ewick, Patricia, and Susan S. Silbey. *The Common Place of Law: Stories from Everyday Life.* Chicago: University of Chicago Press, 1998.

Excelsior. "Segob emite decreto de Ley de Geolocalización o monitoreo celular." April 17, 2012.

Feeley, Malcolm M., and Jonathan Simon. "The New Penology: Notes on the Emerging Strategy of Corrections and Its Implications." *Criminology* 30, no. 4 (November 1, 1992): 449–74.

Fernández Christlieb, Federico. *Las modernas ruedas de la destrucción: El automóvil en la Ciudad de México.* Mexico City: Ediciones el Caballito, 1992.

Fierro, Miguel. "Llevará angulo fraude fiscal del REPUVE a la Cámara de Diputados." *Puente Libre,* April 12, 2013.

Firmino, Rodrigo José, Fernanda Bruno, and Nelson Arteaga Botello. "Understanding the Sociotechnical Networks of Surveillance Practices in Latin America." *Surveillance and Society* 10, no. 1 (July 2012): 1–4.

Flores, Pepe. "La Ley de Geolocalización entra en vigor en México." *Hipertextual,* April 18, 2012.

Fortin, Jacey. "Nigerian Citizens Call for Transparency amid Suspicions of Shady Government Deal for Internet Surveillance." *International Business Times,* June 7, 2013.

Foucault, Michel. *Discipline and Punish: The Birth of the Prison.* New York: Pantheon, 1977.

———. *Security, Territory, Population: Lectures at the Collège de France 1977–1978.* Translated by Graham Burchell. New York: Picador, 2009.

Freeman, Laurie, and Jorge Luis Sierra. "Mexico: The Militarization Trap." In *Drugs and Democracy in Latin America,* edited by Coletta Youngers and Eileen Rosin, 263–302. Boulder, CO: Lynne Rienner, 2005.

Funetes, Víctor. "Buscan poseer la marca Repuve." *Agencia reforma,* April 20, 2013.

García, Dennis. "Alertan por niveles de impunidad en México." *El Universal,* April 21, 2015.

García Martinez, Bernardo. *The Highway of Mexico (1891–1991)*. Mexico City: Grupo Azabache, 1992.

———. "Los años de expansión." In *Nueva historia general de México,* edited by Erik Velásquez García et al., 217–62. Mexico City: Colegio de México, 2010.

———. "Los años de la conquista." In *Nueva historia general de México,* edited by Erik Velásquez García et al., 169–215. Mexico City: Colegio de México, 2010.

Gates, Kelly. "Biometrics and Post-9/11 Technostalgia." *Social Text* 23, no. 2 83 (June 2005): 35–53.

———. "Identifying the 9/11 'Faces of Terror.'" *Cultural Studies* 20, nos. 4–5 (2006): 417–40.

———. *Our Biometric Future: Facial Recognition Technology and the Culture of Surveillance*. New York: New York University Press, 2011.

Geser, Hans. "Is the Cell Phone Undermining the Social Order? Understanding Mobile Technology from a Sociological Perspective." *Knowledge, Technology and Policy* 19, no. 1 (2006): 8–18.

Gilliom, John. *Overseers of the Poor: Surveillance, Resistance, and the Limits of Privacy*. Chicago: University of Chicago Press, 2001.

———. "Struggling with Surveillance: Resistance, Consciousness, and Identity." In *The New Politics of Surveillance and Visibility,* edited by Richard V. Ericson and Kevin D. Haggerty, 111–40. Toronto: University of Toronto Press, 2006.

Gilliom, John, and Torin Monahan. *SuperVision: An Introduction to the Surveillance Society*. Chicago: University of Chicago Press, 2012.

Glanz, James, Sebastian Rotella, and David Sanger. "In 2008 Mumbai Attacks, Piles of Spy Data, but an Uncompleted Puzzle." *New York Times,* December 21, 2014.

Gómez, Ricardo, and Elena Michel. "Comisiones del Senado avalan Ley de Seguridad Nacional." *El Universal,* April 23, 2010.

Gonzales, Michael J. *The Mexican Revolution, 1910–1940*. Albuquerque: University of New Mexico Press, 2002.

Goodin, Dan. "NSA Repeatedly Tries to Unpeel Tor Anonymity and Spy on Users, Memos Show." *Ars Technica,* October 4, 2013.

Goodwin, Jeff, James M. Jasper, and Francesca Polletta. *Passionate Politics: Emotions and Social Movements*. Chicago: University of Chicago Press, 2001.

Goold, Benjamin. "Technologies of Surveillance and the Erosion of Institutional Trust." In *Technologies of InSecurity: The Surveillance of Everyday Life,* edited by Katja Franko Aas, Helene Oppen Gundhus, and Heidi Mork Lomell, 207–18. New York: Routledge-Cavendish, 2008.

Gorlier, Juan Carlos, and Keith Guzik. *La política de género en América Latina: Debates, teorías, metodologías y estudios de caso*. Buenos Aires: Ediciones al Margen, 2002.

Graham, Stephen, and David Wood. "Digitizing Surveillance: Categorization, Space, Inequality." *Critical Social Policy* 23, no. 2 (May 1, 2003): 227–48.

Guadarrama, José. "El Renaut dio como resultado un fracaso millonario." *Excelsior,* August 8, 2011.

———. "Renaut alterno ha dado de baja 12 mil 200 celulares." *Excelsior,* August 6, 2011.

Gusfield, Joseph R. *The Culture of Public Problems: Drinking-Driving and the Symbolic Order.* Chicago: University of Chicago Press, 1984.

Guzik, Keith. "Discrimination by Design: Data Mining in the United States' 'War on Terrorism.'" *Surveillance and Society* 7, no. 1 (October 13, 2009): 3–20.

———. "Security a la Mexicana: On the Particularities of Security Governance in México's War on Crime." *Theory and Society* 42, no. 2 (March 12, 2013): 161–87.

Guzik, Keith, and Juan Carlos Gorlier. "History in the Making: Narrative as Feminist Text and Practice in a Mexican Feminist Journal." *Social Movement Studies* 3, no. 1 (2004): 89–116.

Guzmán Roque, Sharenii. "El 'Bunker' más grande de AL es inaugurado por GDF." *El Universal,* October 25, 2011.

Hacking, Ian. *The Taming of Chance.* Vol. 17. Cambridge: Cambridge University Press, 1990.

Haggerty, Kevin D. "From Risk to Precaution: The Rationalities of Personal Crime Prevention." In *Risk and Morality,* edited by Richard Victor Ericson and Aaron Doyle, 193–214. Toronto: University of Toronto Press, 2003.

———. "Visible War: Surveillance, Speed, and Information War." In *The New Politics of Surveillance and Visibility,* edited by Richard V. Ericson and Kevin D. Haggerty. Toronto: University of Toronto Press, 2006.

Haggerty, Kevin D., and Richard V. Ericson. "The New Politics of Surveillance and Visibility." In *The New Politics of Surveillance and Visibility,* edited by Richard V. Ericson and Kevin D. Haggerty, 3–34. Toronto: University of Toronto Press, 2006.

———. "The Surveillant Assemblage." *British Journal of Sociology* 51, no. 4 (December 1, 2000): 605–22.

Hausberger, Bernd, and Óscar Mazín. "Nueva España: Los años de autonomía." In *Nueva historia general de México,* edited by Erik Velásquez García et al., 263–306. Mexico City: Colegio de México, 2010.

Hernández, Anabel. *Narcoland: The Mexican Drug Lords and Their Godfathers.* Brooklyn: Verso, 2013.

Hernández, Aura. "Facilitarán el bloqueo de celulares robados para frenar extorsión." *La Razón,* November 5, 2010.

Hernández Chávez, Alicia. "El estado nacionalista, su referente histórico." In *The Evolution of the Mexican Political System,* edited by Jaime E. Rodríguez, 203–13. Wilmington, DE: Scholarly Resources, 1993.

Hernández Serrano, Federico. *. . . y se formaron caminos.* Mexico City: Chrysler de México, 1986.

Herrera, Paty. "Se quedan a dormir para poder sacar engomado de REPUVE." *Puente libre,* December 29, 2013.

Herrera Madrano, Daisy. "Circulan 10 mil autos robados en Tamaulipas." *Centro noticias Tamaulipas,* October 25, 2010.

Hickey, Eric W., ed. *Encyclopedia of Murder and Violent Crime.* Thousand Oaks, CA: Sage, 2003.

Holguín, Elena. "Piden condonar los adeudos en 'plaqueo.'" *El Siglo de Torreón,* March 24, 2012.

Hollander, Jocelyn A., and Rachel L. Einwohner. "Conceptualizing Resistance." *Sociological Forum* 19, no. 4 (2004): 533–54.

Hookway, James. "Thailand's Dummy Cops Are Really Watching This Time." *Wall Street Journal,* December 13, 2011.

Horn, Rebecca. "Gender and Social Identity: Nahua Naming Patterns in Postconquest Central Mexico." In *Indian Women of Early Mexico,* edited by Susan Schroeder, Stephanie Wood, and Robert Haskett, 105–22. Norman: University of Oklahoma Press, 1997.

Hoy, Teresa M. Van. "La Marcha Violenta? Railroads and Land in 19th-Century Mexico." *Bulletin of Latin American Research* 19, no. 1 (January 1, 2000): 33–61.

Incháustegui Romer, Teresa, et al. *Violencia feminicida en México: Características, tendencias y nuevas expresiones en las entidades federativas, 1985–2010.* New York: ONU Mujeres, 2012.

Instituto Nacional de Estadística y Geografía. "Encuesta nacional de victimización y percepción sobre seguridad pública 2014." September 30, 2014. www.inegi.org.mx/est/contenidos/proyectos/encuestas/hogares/regulares/envipe/envipe2014/default.aspx.

Jang, Hyunseok, Larry T. Hoover, and Hee-Jong Joo. "An Evaluation of Compstat's Effect on Crime: The Fort Worth Experience." *Police Quarterly* 13, no. 4 (December 1, 2010): 387–412.

Jasanoff, Sheila, ed. *States of Knowledge: The Co-production of Science and the Social Order.* London: Routledge, 2004.

Jenkins, J. Craig. "Resource Mobilization Theory and the Study of Social Movements." *Annual Review of Sociology* 9 (1983): 527–53.

Jones, Toby Craig. *Desert Kingdom How Oil and Water Forged Modern Saudi Arabia.* Cambridge, MA: Harvard University Press, 2010.

Joseph, Gilbert M., and Daniel Nugent. "Popular Culture and State Formation in Revolutionary Mexico." In *Everyday Forms of State Formation: Revolution and the Negotiation of Rule in Modern Mexico,* 3–24. Durham, NC: Duke University Press, 1994.

Juárez, Geraldine. "EFF considera 'alarmante' la nueva Ley de Geolocalización mexicana." *Hipertextual,* March 7, 2012.

Juárez Alfaro, Javier. "Rechazan el Repuve." *El Diario de Ciudad Victoria,* August 20, 2011.

Jusidman, Clara. "El padrón electoral en el camino de la democracia." In *El camino de la democracia en México,* edited by Patricia Galeana, 253–73. Mexico City: Instituto de Investigaciones Jurídicas, 1998.

Keating, Joshua. "No One in Europe Is Tougher on Terror Than France; That Didn't Stop the Attacks." *Slate,* January 13, 2015.

Knight, Alan. "The Modern Mexican State: Theory and Practice." In *The Other Mirror: Grand Theory through the Lens of Latin America,* 177–218. Princeton, NJ: Princeton University Press, 2001.

Koonings, Kees. "New Violence, Insecurity, and the State." In *Violence, Coercion, and State-Making in Twentieth-Century Mexico: The Other Half of the Centaur,* edited by Wil Pansters, 255–78. Stanford, CA: Stanford University Press, 2012.

Korte, Gregory. "Obama Bans Some Military Equipment Sales to Police." *USA Today,* May 18, 2015.

Kuntz Ficker, Sandra, and Elisa Speckman Guerra. "El Porfiriato." In *Nueva historia general de México,* edited by edited by Erik Velásquez García et al., 487–536. Mexico City: Colegio de México, 2010.

La Cultura Jurídica. "El orden constitucional y la banda del *Automovil gris.*" March 6, 2012.

La Jornada. "Construirá el gobierno un búnker secreto para labores de inteligencia antinarcóticos." March 17, 2011.

Latour, Bruno. "Give Me a Lab and I Will Raise the World." In *Science Observed: Perspectives on the Social Studies of Science,* edited by Karen Knorr-Cetina and Michael Mulkay, 141–70. London: Sage, 1983.

———. *Pandora's Hope: Essays on the Reality of Science Studies.* Cambridge, MA: Harvard University Press, 1999.

———. "Where Are the Missing Masses? The Sociology of a Few Mundane Artifacts." In *Shaping Technology/Building Society,* edited by Wiebe Bijker and John Law, 225–58. Cambridge, MA: MIT Press, 1992.

Lee, Ronald D., and Paul M. Schwartz. "Beyond the 'War' on Terrorism: Towards the New Intelligence Network." *Michigan Law Review* 103, no. 6 (May 1, 2005): 1446–82.

Leo, Richard A., and Kimberly D. Richman. "Mandate the Electronic Recording of Police Interrogations." *Criminology and Public Policy* 6, no. 4 (2007): 791–98.

Lévi-Strauss, Claude. *The Savage Mind.* Chicago: University of Chicago Press, 1966.

Levy, Daniel C., Kathleen Bruhn, and Emilio Zebadúa. *Mexico: The Struggle for Democratic Development.* Berkeley: University of California Press, 2001.

Ley del Registro Público Vehicular. 2004. http://cgservicios.df.gob.mx/prontuario/vigente/r068201.htm.

Lin, Chung-hsi. "The Silenced Technology—the Beauty and Sorrow of Reassembled Cars." *East Asian Science, Technology and Society: An International Journal* 3, no. 1 (June 17, 2009): 91–131.

Ling, Rich. *The Mobile Connection: The Cell Phone's Impact on Society.* San Francisco: Morgan Kaufmann, 2004.

Littlefield, Melissa. "Constructing the Organ of Deceit: The Rhetoric of fMRI and Brain Fingerprinting in Post-9/11 America." *Science, Technology and Human Values* 34, no. 3 (May 1, 2009): 365–92.

Lopez, German. "Why Police Should Wear Body Cameras—and Why They Shouldn't." *Vox,* January 13, 2015.

López Galicia, Arely. "Imponen otro cobro a importadores de autos." *Conexión Total,* August 12, 2011.

Lutz, Ellen, and Naomi Roht-Arriaza. "The Cavallo Case: One Year Later." Chapter 13 in *ACLU International Civil Liberties Report*. New York: ACLU, 2003.

Lyon, David. *Identifying Citizens: ID Cards as Surveillance*. Malden, MA: Polity, 2009.

———. "National IDs in a Global World: Surveillance, Security, and Citizenship." *Case Western Reserve Journal of International Law* 42, no. 3 (2010): 607–23.

———. *Surveillance after September 11*. Malden, MA: Polity, 2003.

———. *Surveillance Society: Monitoring Everyday Life*. Buckingham, UK: Open University Press, 2001.

MacLachlan, Colin M. *Criminal Justice in Eighteenth Century Mexico: A Study of the Tribunal of the Acordada*. Berkeley: University of California Press, 1974.

Mann, Steve, Jason Nolan, and Barry Wellman. "Sousveillance: Inventing and Using Wearable Computing Devices for Data Collection in Surveillance Environments." *Surveillance and Society* 1, no. 3 (2002): 331–55.

Manning, Peter K. *The Technology of Policing: Crime Mapping, Information Technology, and the Rationality of Crime Control*. New York: New York University Press, 2008.

Marcuse, Herbert. *One-Dimensional Man: Studies in the Ideology of Advanced Industrial Society*. Boston: Beacon Press, 1964.

Markon, Jerry, and Ellen Nakashima. "NSA Director Says Surveillance Programs Thwarted 'Dozens' of Attacks." *Washington Post*, June 12, 2013.

Márquez, Graciela, and Lorenzo Meyer. "Del autoritarismo agotado a la democracia frágil." In *Nueva historia general de México*, edited by Erik Velásquez García et al., 747–91. Mexico City: Colegio de México, 2010.

Marres, Noortje, and Javier Lezaun. "Materials and Devices of the Public: An Introduction." *Economy and Society* 40, no. 4 (November 1, 2011): 489–509.

Martínez, Fernando, and Alberto Morales. "Exige Martí que gobiernos se coordinen contra secuestros." *El Universal*, August 13, 2008.

Martínez, Víctor. "Instaló finanzas 8 mil chips del Repuve." *Ntrzacatecas.com*, January 16, 2013.

Marx, Gary T. "Seeing Hazily (But Not Darkly) through the Lens: Some Recent Empirical Studies of Surveillance Technologies." *Law and Social Inquiry* 30, no. 2 (2005): 339–99.

———. "Surveillance and Society." In *Encyclopedia of Social Theory*, edited by George Ritzer, 817–22. Thousand Oaks, CA: Sage, 2005.

———. "Technology and Social Control." In *International Encyclopedia of the Social and Behavioral Sciences*, edited by Neil J. Smelser and Paul B. Baltes. Philadelphia: Elsevier, 2001. Accessed June 19, 2014. http://web.mit.edu/gtmarx/www/techandsocial.html.

———. *Windows into the Soul*. Chicago: University of Chicago Press, 2016.

Matute Aguirre, Alvaro. "La encrucijada de 1929: Caudillismo versus institucionalización." In *The Evolution of the Mexican Political System*, edited by Jaime E. Rodríguez O., 187–202. Wilmington, DE: Scholarly Resources, 1993.

McAdam, Doug, John D. McCarthy, and Mayer N. Zald. *Comparative Perspectives on Social Movements: Political Opportunities, Mobilizing Structures, and Cultural Framings*. Cambridge: Cambridge University Press, 1996.

McCann, Michael W. *Rights at Work: Pay Equity Reform and the Politics of Legal Mobilization*. Chicago: University of Chicago Press, 1994.

McShane, Clay. "The Origins and Globalization of Traffic Control Signals." *Journal of Urban History* 25, no. 3 (March 1, 1999): 379–404.

Medieta, Susana. "Exige el IFAI destruir los datos reunidos con el derogado Renaut." *Milenio*, March 3, 2012.

Medina, Eden. *Cybernetic Revolutionaries Technology and Politics in Allende's Chile*. Cambridge, MA: MIT Press, 2011.

Mejía, José Gerardo. "Empresa extranjera hará cédula." *El Universal*, November 25, 2009.

———. "La corte analiza queja por cédula." *El Universal*, March 7, 2011.

Melucci, Alberto. *Nomads of the Present: Social Movements and Individual Needs in Contemporary Society*. Philadelphia: Temple University Press, 1989.

Mendoza, Arturo. "Comparecen los titulares de SCT, Cofetel, y Cofeco ante Comisión de Comunicación." *El Punto Crítico*, March 31, 2011.

Merry, Sally Engle. "Resistance and the Cultural Power of Law." *Law and Society Review* 29, no. 1 (1995): 11–27.

Meserve, Jeanne, and Mike Ahlers. "Drone Crash in El Paso under Investigation." *CNN.com*, December 17, 2010.

Michel, Elena, and Ricardo Gómez. "Avanza ley contra secuestro." *El Universal*, March 26, 2010.

Milenio. "Entrega Calderón primeras Cédulas de Identidad Personal para Menores." March 15, 2011.

———. "Equivocada, iniciativa de Peña sobre telecom." April 23, 2014.

———. "Inauguran centro de mando en Huixquilucan." August 4, 2010.

———. "Telcel y Movistar desactivarán teléfonos celulares robados." April 20, 2010.

Molotch, Harvey. *Against Security: How We Go Wrong at Airports, Subways, and Other Sites of Ambiguous Danger*. Princeton, NJ: Princeton University Press, 2012.

Molyneux, Maxine. "Mobilization without Emancipation? Women's Interests, the State, and Revolution in Nicaragua." *Feminist Studies* 11, no. 2 (1985): 227–54.

Monahan, Torin. "Electronic Fortification in Phoenix Surveillance Technologies and Social Regulation in Residential Communities." *Urban Affairs Review* 42, no. 2 (November 1, 2006): 169–92.

———. "Surveillance as Governance: Social Inequality and the Pursuit of Democratic Surveillance." In *Surveillance and Democracy*, edited by Kevin D. Haggerty and Minas Samatas, 91–110. New York: Routledge-Cavendish, 2010.

———. *Surveillance in the Time of Insecurity*. Piscataway, NJ: Rutgers University Press, 2010.

———. "'War Rooms' of the Street: Surveillance Practices in Transportation Control Centers." *Communication Review* 10, no. 4 (2007): 367–89.

Monahan, Torin, and Tyler Wall. "Somatic Surveillance: Corporeal Control through Information Networks." *Surveillance and Society* 4, no. 3 (September 1, 2002): 154–73.

Monroy, Jorge. "Extorsiones aumentan 40% por Renaut." *El Economista,* February 13, 2011.

———. "Nuevo diseño de la credencial del IFE encriptará datos." *El Economista,* November 13, 2012.

Moore, Dawn. *Criminal Artefacts Governing Drugs and Users.* Vancouver: University of British Columbia Press, 2008.

Morales Zebadúa, Carlos Antonio. "Breve historia de pifias en la protección de los datos personales en México." *El Universal,* April 21, 2010.

Morris, Stephen D. *Political Corruption in Mexico: The Impact of Democratization.* Boulder, CO: Lynne Rienner, 2009.

Moser, Caroline O. N. "Editor's Introduction: Urban Violence and Insecurity; An Introductory Roadmap." *Environment and Urbanization* 16, no. 2 (2004): 3.

Müller, Markus-Michael. *Public Security in the Negotiated State: Policing in Latin America and Beyond.* New York: Palgrave Macmillan, 2012.

Mumford, Lewis. "Technics and the Nature of Man." *Technology and Culture* 7, no. 3 (1966): 303–17.

Mundodehoy.com. "Impuesto a coches importados." August 13, 2011.

Murakami Wood, David. "Cameras in Context: A Comparison of the Place of Video Surveillance in Japan and Brazil." In *Eyes Everywhere: The Global Growth of Camera Surveillance,* edited by Aaron Doyle, Randy Lippert, and David Lyon, 83–99. Abingdon, UK: Routledge, 2011.

———. "Globalization and Surveillance." In *Routledge Handbook of Surveillance Studies,* edited by Kirstie Ball, Kevin D. Haggerty, and David Lyon, 333–42. Abingdon, UK: Routledge, 2012.

———. "What Is Global Surveillance? Towards a Relational Political Economy of the Global Surveillant Assemblage." *Geoforum* 49 (October 2013): 317–26.

Nacif Hernández, Benito. "El INE frente al robo de identidad." *El Universal,* August 13, 2015.

Navitski, Rielle. "Spectacles of Violence and Politics: *El Automóvil Gris* (1919) and Revolutionary Mexico's Sensational Visual Culture." *Journal of Latin American Cultural Studies* 23, no. 2 (2014): 133–52.

Nellis, Mike. "Mobility, Locatability and the Satellite Tracking of Offenders." In *Technologies of InSecurity: The Surveillance of Everyday Life,* edited by Katja Franko Aas, Helene Oppen Gundhus, and Heidi Mork Lomell, 105–24. New York: Routledge-Cavendish, 2008.

Nelson Reames, Benjamin. "A Profile of Police Forces in Mexico." In *Reforming the Administration of Justice in Mexico,* edited by Wayne A. Cornelius and David Shirk, 117–32. Notre Dame, IN: University of Notre Dame Press, 2007.

Neyland, Daniel. "Mundane Terror and the Threat of Everyday Objects." In *Technologies of InSecurity: The Surveillance of Everyday Life,* edited by Katja Franko Aas, Helene Oppen Gundhus, and Heidi Mork Lomell, 21–41. New York: Routledge-Cavendish, 2008.

Norris, Clive. "The Success of Failure: Accounting for the Global Growth of CCTV." In *Routledge Handbook of Surveillance Studies,* edited by Kirstie Ball, Kevin D. Haggerty, and David Lyon, 251–58. New York: Routledge, 2012.

Norris, Clive, and Gary Armstrong. *The Maximum Surveillance Society: The Rise of CCTV.* Oxford: Bloomsbury Academic, 1999.

Notimex. "Aprueba Congreso de SLP que REPUVE sea obligatorio hasta 2014." January 13, 2013.

———. "Detallan beneficios de la Cédula de Identidad personal." January 13, 2011.

———. "Se alista el sector automotriz para negociar con nueva legislatura." July 7, 2009.

———. "Titular de Cofetel defiende 'éxito' de Renaut." April 20, 2010.

Nunn, Samuel. "Cities, Space, and the New World of Urban Law Enforcement Technologies." *Journal of Urban Affairs* 23, nos. 3–4 (July 2001): 259–78.

O'Connor, Anne-Marie. "Mexico City Drastically Reduced Air Pollutants since 1990s." *Washington Post,* April 1, 2010.

O'Donnell, Guillermo. "Reflections on Contemporary South American Democracies." *Journal of Latin American Studies* 33, no. 3 (2001): 599–609.

Olivares Alonso, Emir. "Concretar la Cédula de Identidad significa echar a la basura $40 mil millones, advierten." *La Jornada,* August 28, 2009.

Olson, Mancur. *The Logic of Collective Action: Public Goods and the Theory of Groups.* Cambridge, MA: Harvard University Press, 1965.

Olvera, Carlos. "Población desconfía de los registros." *Periódico correo,* August 2, 2011.

Online Etymology Dictionary. Accessed January 11, 2015. www.etymonline.com.

Organisation for Economic Co-operation and Development. *Collusion and Corruption in Public Procurement.* Paris: OECD, October 15, 2010.

Ortega, Eduardo. "Gendarmería Nacional será una policía cercana a la gente: Peña Nieto." *El Financiero,* August 22, 2014.

Pacto por México. Accessed February 6, 2015. http://pactopormexico.org.

Pansters, Wil. "Zones of State-Making: Violence, Coercion, and Hegemony in Twentieth-Century Mexico." In *Violence, Coercion, and State-Making in Twentieth-Century Mexico: The Other Half of the Centaur,* edited by Wil Pansters, 3–42. Stanford, CA: Stanford University Press, 2012.

Passoth, Jan-Hendrik, and Nicholas Rowland. "Actor-Network State." *International Sociology* 25, no. 6 (2010): 818–41.

Pereira, Anthony W., and Diane E. Davis. "New Patterns of Militarized Violence and Coercion in the Americas." *Latin American Perspectives* 27, no. 2 (March 1, 2000): 3–17.

Pickering, Andrew. *The Mangle of Practice: Time, Agency, and Science.* Chicago: University of Chicago Press, 1995.

———. "New Ontologies." In *The Mangle in Practice: Science, Society, and Becoming,* edited by Andrew Pickering and Keith Guzik, 1–14. Durham, NC: Duke University Press, 2008.

Pickler, Nedra. "Experts Say Drone Strikes Appear in Bounds of US Law." *Yahoo News.* Accessed August 10, 2015. http://news.yahoo.com/experts-drone-strikes-appear-bounds-us-law-191859227—politics.html.

Pinch, Trevor, and Wiebe Bijker. "The Social Construction of Facts and Artefacts; or, How the Sociology of Science and the Sociology of Technology Might Benefit Each Other." *Social Studies of Science* 14, no. 3 (1984): 399–441.

Plana, Manuel. *Las industrias, siglos XVI al XX.* Mexico City: Oceano de Mexico, 2004.

Poiré Romero, Alejandro. "It Doesn't Come Easy—Mexico's Fight for Security." Keynote address at INFO Summit 2012. www.youtube.com/watch?v=Q5Gck2iwnbs&feature=youtu.be.

Polgreen, Lydia. "With National Database, India Tries to Reach the Poor." *New York Times,* September 1, 2011.

Polletta, Francesca. "'It Was Like a Fever': Narrative and Identity in Social Protest." *Social Problems* 45, no. 2 (1998): 137–59.

Ponce, Diana. "Inicia a nivel nacional el Registro Público Vehicular en Zacatecas." *El Sol de Zacatecas,* July 23, 2010.

Posada García, Miriam. "Movistar perderá la concesión si incumple la ley, advierte Cofetel." *La Jornada,* April 10, 2010.

Poster, Mark. *The Mode of Information: Poststructuralism and Social Context.* Chicago: University of Chicago Press, 1990.

Powell, Kathy. "Political Violence, Everyday Political Violence, and Electoral Processes during the Neoliberal Period in Mexico." In *Violence, Coercion, and State-Making in Twentieth-Century Mexico: The Other Half of the Centaur,* edited by Wil Pansters, 212–32. Stanford, CA: Stanford University Press, 2012.

Prado, Henia. "Presume SSP su plataforma México." *Reforma,* August 12, 2012.

Presidencia de la República. "Diversas intervenciones en el inicio del registro para obtener la Cédula de Identidad Personal: Como tú no hay dos." June 10, 2011. http://calderon.presidencia.gob.mx/2011/06/diversas-intervenciones-en-el-inicio-del-registro-para-obtener-la-cedula-de-identidad-personal-como-tu-no-hay-dos.

———. "Para combatir al crimen, inaugura presidente el Centro de Inteligencia de la Policía Federal." November 24, 2009. http://calderon.presidencia.gob.mx/2009/11/para-combatir-al-crimen-inaugura-presidente-el-centro-de-inteligencia-de-la-policia-federal.

Pulso. "Policías ayudarán a compradores a verificar status de autos." August 16, 2011.

Quiroz, Carlos. "Cédula de Identidad sigue en pie: Secretaría de Gobernación." *Excelsior,* March 6, 2013.

Quiroz, Carlos, and Aurora Zepeda. "Gobernación para expedir la CIC obligó al gobierno federal a diferir el plan." *Excelsior,* January 21, 2011.

Radial Sur. "'Libre tránsito' para gente del sur." September 15, 2011.

Ramírez, Brenda. "Inoperate el registro vehicular." *El Diario de Morelos,* August 16, 2010.

Ramírez de la O, Rogelio. "The Impact of NAFTA on the Auto Industry in Mexico." In *The North American Auto Industry under NAFTA,* edited by Sidney Weintraub and Christopher Sands, 48–91. Washington, DC: CSIS Press, 1998.

Raphael, Ricardo. "El chip de campa." *El Universal,* September 8, 2008.

Rebolledo, Antonio. "Ilegal, cobro de Repuve por Banjército, acusan." *El Diario,* May 14, 2013.

Recillas Enecoiz, Luis. *"El automóvil gris* (1919)." Cine Silente Mexicano / Mexican Silent Cinema. Accessed February 4, 2015. https://cinesilentemexicano.wordpress.com/2009/07/08/el-automovil-gris-1919.

Registro Nacional de Población (RENAPO). "¿Qué es la Cédula de Identidad Personal (registro de menores de edad)?" 2011. www.renapo.gob.mx/CEDI.html.

Replogle, Jill. "Senate Immigration Bill Calls for a Drone-Patrolled Border." *Fronteras Desk,* April 23, 2013.

Reséndiz, Francisco, and Alberto Morales. "EPN: Hasta escépticos ven avances en seguridad." *El Universal,* March 31, 2015.

Reyes, José Juan. "InfoDF pide a la Segob eliminar datos del Renaut." *El Economista,* May 9, 2011.

Ribando Seelke, Clare, and Kristin Finklea. *U.S.-Mexican Security Cooperation: The Mérida Initiative and Beyond.* CRS Report R41349. Washington, DC: Congressional Research Service, February 16, 2011.

Rodríguez, Francisco. "Campa Cifrián—García Luna." *Índice político,* September 5, 2012.

Rodríguez, Martín. "Pernoctan en el parque por el Repuve." *Pulso,* October 28, 2012.

Rodríguez Manzo, Graciela. "Geolocalización en el país de Las Maravillas." *Proceso,* January 16, 2014.

Román, Romina. "Robo de autos creció 86% en seis años: AMIS." *El Universal,* May 18, 2012.

Rose, Nikolas. *Powers of Freedom: Reframing Political Thought.* New York: Cambridge University Press, 1999.

Rose, Nikolas, Pat O'Malley, and Mariana Valverde. "Governmentality." *Annual Review of Law and Social Science* 2, no. 1 (December 2006): 83–104.

Roxborough, Ian. *Unions and Politics in Mexico: The Case of the Automobile Industry.* New York: Cambridge University Press, 1984.

Rubenstein, Anne. "Mass Media and Popular Culture in the Postrevolutionary Era." In *The Oxford History of Mexico,* edited by William Beezley and Michael Meyer, 598–633. New York: Oxford University Press, 2010.

Rubin, Allen, and Earl R. Babbie. *Essential Research Methods for Social Work.* Belmont, CA: Thomson/Brooks/Cole, 2007.

Sabet, Daniel. *Police Reform in Mexico: Informal Politics and the Challenge of Institutional Change.* Stanford, CA: Stanford Politics and Policy, 2012.

Sánchez, Emma. "Arranca Registro Público Vehicular en Michoacán." *Provincia.* Accessed February 6, 2015. www.provincia.com.mx/arranca-registro-publico-vehicular-en-michoacan.

Sánchez Ley, Laura. "Inútil, registro de celulares: Coparmex." *El Sol de Tijuana,* April 12, 2011.

———. "30 millones con datos falsos en el Renaut." *El Sol de Tijuana,* April 12, 2011.

Sanchiz, Santiago. *Manual del "chauffeur."* Mexico City: n.p., 1943.

Sanger, Carol. "Girls and the Getaway: Cars, Culture, and the Predicament of Gendered Space." *University of Pennsylvania Law Review* 144, no. 2 (December 1, 1995): 705–56.

Santillán, Gerardo. "20 mil vehículos dados de alta en el Repuve Tlaxcala." *E-Tlaxcala.mx,* January 15, 2013.

Sarakki Associates Incorporated. *Mexico's National Command and Control Center Challenges and Successes.* June 22, 2011. http://dodccrp.org/events/16th_iccrts_2011/presentations/141.pdf.

Scott, James. *Seeing Like a State: How Certain Schemes to Improve the Human Condition Have Failed.* New Haven, CT: Yale University Press, 1998.

———. *Weapons of the Weak: Everyday Forms of Peasant Resistance.* New Haven, CT: Yale University Press, 1985.

Secretaría de Comunicaciones y Obras Públicas (SCOP). *Seguridad! El transito en las carreteras federales. Disposiciones reglamentarias. Accidentes. Infracciones. Sanciones.* Mexico City: SCOP, 1938.

Secretaría de Comunicaciones y Transportes. "Decreto por el que se reforman y adicionan diversas disposiciones de la Ley Federal de Telecomunicaciones." February 9, 2009. www.renaut.gob.mx/work/models/RENAUT/Resource/328/1/images/decreto.pdf.

Secretaría de Gobernación (SEGOB). "Inicio del registro para obtener la Cédula de Identidad para Menores." 2011. www.youtube.com/watch?v=rbVn1-L42sw&feature=relmfu.

———. "Preguntas frecuentes." 2011. www.renaut.gob.mx/en/RENAUT/Preguntas_frecuentes1.

Secretaría de Medio Ambiente. *Elementos para la propuesta de actualización del programa "Hoy No Circula" de la Zona Metropolitana del Valle de México.* Mexico City: Gobierno del Distrito Federal—Dirección General de Gestión Ambiental del Aire, 2004.

Secretaría de Seguridad Pública. "Sobre REPUVE." 2009. www.repuve.gob.mx/acerca.html.

Semanario ZETA. "Los muertos de EPN: 36 mil 718." August 28, 2014.

Serrano, Mónica. "States of Violence: State-Crime Relations in Mexico." In *Violence, Coercion, and State-Making in Twentieth-Century Mexico: The Other Half of the Centaur,* edited by Wil Pansters, 135–58. Stanford, CA: Stanford University Press, 2012.

Serrano, Mónica, and María Celia Toro. "Del narcotráfico al crimen transnacional organizado en América Latina." In *Crimen transnacional organizado y seguridad internacional,* edited by Mats Berdal, 233–73. Mexico City: Fondo de Cultura Económica, 2005.

Serrano Ortega, José Antonio, and Josefina Zoraida Vázquez. "El nuevo orden: 1821–1848." In *Nueva historia general de México,* edited by Erik Velásquez García et al., 397–442. Mexico City: Colegio de México, 2010.

Sewell, William H. "A Theory of Structure: Duality, Agency, and Transformation." *American Journal of Sociology* 98, no. 1 (1992): 1–29.

Shallwani, Pervaiz. "NYPD Unveil Two Cameras for Officers." *Wall Street Journal,* September 5, 2014.

Shapin, Steven. "Pump and Circumstance: Robert Boyle's Literary Technology." *Social Studies of Science* 14, no. 4 (1984): 481–520.

Shirk, David A., and Alejandra Ríos Cázares. "Introduction: Reforming the Administration of Justice in Mexico." In *Reforming the Administration of Justice in Mexico,* edited by Wayne A. Cornelius and David A. Shirk, 1–47. Notre Dame, IN: University of Notre Dame Press, 2007.

Silverman, Eli. "Compstat's Innovation." In *Police Innovation: Contrasting Perspectives,* edited by David Weisburd and Anthony A. Braga, 267–83. Cambridge: Cambridge University Press, 2006.

Simmons, Ric. "Why 2007 Is Not Like 1984: A Broader Perspective on Technology's Effect on Privacy and Fourth Amendment Jurisprudence." *Journal of Criminal Law and Criminology* 97, no. 2 (Winter 2007): 531–68.

Simon, Jonathan. "Driving Governmentality: Automobile Accidents, Insurance, and the Challenge to Social Order in the Inter-War Years, 1919 to 1941." *Connecticut Insurance Law Journal* 4, no. 2 (1997): 521–88.

———. *Governing through Crime: How the War on Crime Transformed American Democracy and Created a Culture of Fear.* Oxford: Oxford University Press, 2009.

Sipse.com. "Registro vehícular abre sus puertas los sábados." April 8, 2014.

Smith, Gavin J. D. "Exploring Relations between Watchers and Watched in Control(led) Systems: Strategies and Tactics." *Surveillance and Society* 4, no. 4 (2007): 280–313.

Soliloquio21. "Trucos para evadir RENAUT." *Soliloquio21.* Accessed September 9, 2014. http://soliloquio21.blogspot.com/2010/04/renaut-trucos-para-evadirlo.html.

Sparks, Matthew. "Fast Capitalism/Slow Terror: Cushy Cosmopolitanism and Its Extraordinary Others." In *Risk and the War on Terror,* edited by Louise Amoore and Marieke de Goede, 133–57. London: Routledge, 2008.

Star-Ledger Editorial Board. "Track Domestic Abusers with GPS Devices: Editorial." *NJ.com,* October 12, 2014.

Steinhauer, Jennifer, and Jonathan Weisman. "U.S. Surveillance in Place since 9/11 Is Sharply Limited." *New York Times,* June 2, 2015.

Sweet, William, and Stephen Cass. "How to Fight Crime in Real Time." *IEEE Spectrum,* June 1, 2007.

Tanck de Estrada, Dorothy, and Carlos Marichal. "¿Reino o colonia? Nueva España, 1750–1804." In *Nueva historia general de México,* edited by Erik Velásquez García, 307–53. Mexico City: Colegio de México, 2010.

Terra. "Policía Federal muestra interior del Centro de Inteligencia." Accessed August 16, 2015. http://noticias.terra.com.mx/mexico/seguridad/policia-federal-muestra-interior-del-centro-de-inteligencia,84d0ee81b2fd8310Vg nVCM20000099cceb0aRCRD.html.

Thompson, Ginger, and Mark Mazzetti. "U.S. Sends Drones to Fight Mexican Drug Trade." *New York Times,* March 15, 2011.

244 | Bibliography

Tilly, Charles. *Coercion, Capital and European States, AD 990–1992.* Malden, MA: Blackwell, 1990.

Toobin, Jeffrey. "The Solace of Oblivion." *New Yorker,* September 29, 2014.

Toro, Maria Celia, and Mónica Serrano. "From Drug Trafficking to Transnational Organized Crime in Latin America." In *Transnational Organized Crime and International Security: Business as Usual?,* edited by Mats R. Berdal and Mónica Serrano, 155–82. Boulder, CO: Lynne Rienner, 2002.

Touraine, Alain. "An Introduction to the Study of Social Movements." *Social Research* 52, no. 4 (1985): 749–87.

Transparencia Mexicana. "Informe." Accessed February 3, 2015. www.transparenciamexicana.org.mx/Documentos/PactosIntegridad/INFORME_LICITACION_ADQUISICION_EQUIPO_BIOMETRICA.pdf.

Urry, John. "The 'System'of Automobility." *Theory, Culture and Society* 21, nos. 4–5 (2004): 25–39.

Valenzuela Zuñiga, Carolina, Anamaría Silva Mena, Paulina Vargas Herrera, and Mario Vergara Barahona. *Technological Innovation for Security in Latin America.* Santiago: Universidad de Santiago de Chile, 2014.

Valverde, Mariana. "Police Science, British Style: Pub Licensing and Knowledges of Urban Disorder." *Economy and Society* 32, no. 2 (2003): 234–52.

———. "Targeted Governance and the Problem of Desire." In *Risk and Morality,* edited by Richard Victor Ericson and Aaron Doyle, 438–58. Toronto: University of Toronto Press, 2003.

Vanderwood, Paul J. *Disorder and Progress: Bandits, Police, and Mexican Development.* Wilmington, DE: Scholarly Resources, 1992.

van Dijk, J., J. van Kesteren, and Paul Smit. *Criminal Victimisation in International Perspective.* The Hague: Boom Juridische Uitgevers, 2007.

Van Hoy, Teresa. "La Marcha Violenta? Railroads and Land in 19th-Century Mexico." *Bulletin of Latin American Research* 19, no. 1 (January 1, 2000): 33–61.

———. *A Social History of Mexico's Railroads: Peons, Prisoners, and Priests.* Lanham, MD: Rowman and Littlefield, 2008.

Vargas, Jorge A. "Privacy Rights under Mexican Law: Emergence and Legal Configuration of a Panoply of New Rights." *Houston Journal of International Law* 27, no. 1 (2004): 73–136.

Velasco, Jose Luis. *Insurgency, Authoritarianism, and Drug Trafficking in Mexico's "Democratization."* New York: Routledge, 2005.

Vélez, Alejandro. "Insecure Identities: The Approval of a Biometric ID Card in Mexico." *Surveillance and Society* 10, no. 1 (July 18, 2012): 42–50.

Veracruzanos.info. "Veracruz, primer lugar en registro vehicular." Accessed September 14, 2015. www.veracruzanos.info/veracruz-primer-lugar-en-registro-vehicular.

Verrips, Jojada, and Birgit Meyer. "Kwaku's Car: The Struggles and Stories of a Ghanaian Long-Distance Taxi-Driver." In *Car Cultures,* edited by Daniel Miller, 153–65. Oxford: Berg Publishers, 2008.

Villamil, Jenaro. "Atenco, Ibero y la primavera mexicana en 2012." *Proceso,* May 15, 2012.

Villanueva, Ernesto. "Mi columna semanal." *Ernesto Villanueva,* April 2010. http://ernestovillanueva.blogspot.com/2010_04_01_archive.html.

Virilio, Paul. *The Information Bomb.* London: Verso, 2006.

Warren, Richard A. "Mass Mobilization versus Social Control: Vagrancy and Political Order in Early Republican Mexico." In *Reconstructing Criminality in Latin America,* edited by Carlos Aguirre and Robert Buffington, 41–58. Wilmington, DE: Scholarly Resources, 2000.

Washington State Department of Transportation. "List of Vehicles with Metallized Windshields." Accessed February 5, 2015. www.wsdot. wa.gov/NR/rdonlyres/77735B79–209C–4975–8FB5–69DC1B35C1B2/0/ Metallized_Windshield_List_2013.pdf.

Whitaker, Reg. "A Faustian Bargain? America and the Dream of Total Information Awareness." In *The New Politics of Surveillance and Visibility,* edited by Richard V. Ericson and Kevin D. Haggerty, 141–70. Toronto: University of Toronto Press, 2006.

Willis, James J., Stephen D. Mastrofski, and David Weisburd. "Making Sense of COMPSTAT: A Theory-Based Analysis of Organizational Change in Three Police Departments." *Law and Society Review* 41, no. 1 (March 1, 2007): 147–88.

Wilson, Scott. "Road Pricing: Brazil to Have Compulsory Toll Tags by July 2014 (SINIAV)." *Road Pricing,* August 14, 2012. http://roadpricing. blogspot.com/2012/08/brazil-to-have-compulsory-toll-tags-by.html.

Womack, James P. *The Mexican Motor Industry: Strategies for the 1990s.* Cambridge, MA: International Motor Vehicle Program, Massachusetts Institute of Technology, 1989.

Wood, David, Eli Konvitz, and Kirstie Ball. "The Constant State of Emergency? Surveillance after 9/11." In *The Intensification of Surveillance: Crime, Terrorism and Warfare in the Information Age,* edited by Kirstie Ball and Frank Webster, 137–50. London: Pluto Press, 2003.

World Health Organization. *Global Status Report on Road Safety.* Geneva: World Health Organization, 2009. http://whqlibdoc.who.int/ publications/2009/9789241563840_eng.pdf.

Yin, Robert K. *Case Study Research: Design and Methods.* 5th ed. Los Angeles: Sage, 2013.

Zárate, Luz. "Desconocen beneficios de Cédula de Identidad." May 20, 2011. http://anuario.upn.mx/2011/index.php/am-guanajuato/34916-desconocen-beneficios-de-cedula-de-identidad.html.

Zegarra Ballón T, Alberto. *Censo de automóviles en América Latina.* Mexico City: Editorial Ausonia, 1959.

Zureik, Elia, Lynda Harling Stalker, Emily Smith, David Lyon, and Yolande E. Chan, eds. *Surveillance, Privacy, and the Globalization of Personal Information: International Comparisons.* Montreal: McGill-Queen's University Press, 2010.

Index

2016 Vol. 72, No. 3

Proactive Behavior across Group Boundaries: Seeking and Maintaining Positive Interactions with Outgroup Members

Issue Editors: Birte Siem, Stefan Stürmer, and Todd L. Pittinsky